Effective .NET Memory Management

Build memory-efficient cross-platform applications using .NET Core

Trevoir Williams

‹packt›

Effective .NET Memory Management

Group Product Manager: Kunal Sawant
Publishing Product Manager: Teny Thomas
Book Project Manager: Manisha Singh
Senior Editor: Kinnari Chohan
Technical Editor: Vidhisha Patidar
Copy Editor: Safis Editing
Proofreader: Kinnari Chohan
Indexer: Pratik Shirodkar
Production Designer: Jyoti Kadam
DevRel Marketing Coordinator: Sonia Chauhan

First published: July 2024

Production reference: 1120724

Published by Packt Publishing Ltd.
Grosvenor House
11 St Paul's Square
Birmingham
B3 1RB, UK

ISBN 978-1-83546-104-4

www.packtpub.com

I want to express my deepest gratitude to everyone who has supported me throughout this journey.

To my loving wife, your unwavering support and encouragement have been my rock. Your patience, understanding, and belief in me have made all the difference, and I am forever grateful for your presence in my life.

To my parents, thank you for investing in my education. Your sacrifices, guidance, and constant encouragement have shaped me into who I am today. Your belief in the power of education and hard work has been a driving force behind my achievements.

Thank you to my publishing partners for their patience and understanding during the dark days. Your support and dedication have been instrumental in bringing this book to life. I am grateful for your belief in this project and your unwavering commitment to its success.

To my students and colleagues, your enthusiasm and support have been a constant source of inspiration. Your eagerness to learn and collaborate has fueled my passion for teaching and developing innovative solutions. Thank you for being a part of this journey and believing in our shared vision.

This book is a testament to all these incredible individuals' collective efforts and support. I am deeply grateful for every one of you.

Contributors

About the author

Trevoir Williams, a passionate software and system engineer from Jamaica, shares his extensive knowledge with students worldwide. Holding a Master's degree in Computer Science with a focus on Software Development and multiple Microsoft Azure Certifications, his educational background is robust.

His diverse experience includes software consulting, engineering, database development, cloud systems, server administration, and lecturing, reflecting his commitment to technological excellence and education. He is also a talented musician, showcasing his versatility.

He has penned works like *Microservices Design Patterns in .NET* and *Azure Integration Guide for Business*. His practical approach to teaching helps students grasp both theory and real-world applications.

About the reviewer

Panagiotis Malamas is a Microsoft Certified Solution Developer/MCP, a senior full stack software engineer/architect, and an information security analyst. He has 15 years of experience in the IT industry, primarily in the UK, having worked in a plethora of industries and sectors.

He has a rare knack for building fancy PCs and is also a huge motorsports fan.

Table of Contents

8

Performance Considerations and Best Practices 209

9

Preface

In the dynamic world of software development, memory management is a critical yet often overlooked subject that significantly impacts the performance and reliability of applications. As developers, we are constantly seeking ways to optimize our code, reduce resource consumption, and enhance the efficiency of our applications. .NET, with its advanced garbage collection and memory allocation mechanisms, provides a robust environment for building high-performance applications. Still, it also requires a deep understanding of memory management to fully harness its potential.

Effective .NET Memory Management is designed to be your comprehensive guide to mastering the intricacies of memory management within the .NET framework. This book aims to demystify the complexities of garbage collection, memory optimization, and performance tuning, providing you with the knowledge and tools necessary to create efficient, high-performing applications.

.NET has become a key element of modern software development, powering various applications from enterprise-level systems to mobile apps. Its sophisticated memory management features, including automatic garbage collection, make it easier for developers to focus on writing code without worrying about low-level memory allocation and deallocation. However, to truly optimize the performance of .NET applications, a thorough understanding of these memory management mechanisms is essential.

Throughout this book, we will explore the fundamental concepts of .NET memory management, from the basics of the **Common Language Runtime** (**CLR**) and garbage collection to advanced memory profiling and optimization techniques. You will learn how to identify and resolve memory leaks, optimize memory usage, and apply best practices for efficient memory management in your .NET applications.

By the end of this book, you will have a solid understanding of how memory is managed in .NET and the skills to optimize your applications for maximum performance and reliability. Whether you are a novice developer or an experienced professional, *Effective .NET Memory Management* will equip you with the knowledge and expertise needed to take your .NET development skills to the next level.

Join me on this journey to uncover the secrets of .NET memory management and unlock the full potential of your applications. Let's dive in and start optimizing!

Who this book is for

This book is crafted to serve a diverse range of readers involved in .NET development and seek to deepen their understanding of memory management. The target audience includes:

Software engineers: Professionals responsible for designing and implementing software systems will benefit from the advanced techniques and best practices discussed in this book, enabling them to build more robust and efficient applications

System architects: Those involved in the overall design and architecture of software systems will find the detailed exploration of .NET memory management essential for making informed decisions about system design and resource allocation

Technical leads and managers: Leaders who oversee development teams can use this book to ensure their teams follow best practices in memory management, leading to more reliable and high-performance software solutions

Students and educators: Individuals in academic settings, including students pursuing computer science or software engineering degrees and educators teaching .NET technologies, will find this book a valuable resource for learning and teaching memory management concepts

No matter your level of experience or specific role within the software development lifecycle, this book provides the knowledge and tools needed to master the intricacies of memory management in .NET, ultimately leading to the creation of more efficient, reliable, and high-performing applications.

What this book covers

Chapter 1, Memory Management Fundamentals, helps begin our journey into the world of .NET memory management by laying the groundwork with essential concepts and principles. The chapter starts with an overview of the importance of effective memory management in software development, highlighting how it directly impacts application performance and reliability.

Chapter 2, Object Lifetimes and Garbage Collection, dives deeper into the core mechanisms that govern memory management in the .NET framework, focusing on object lifetimes and garbage collection. Understanding how and when objects are created, used, and eventually discarded is crucial for developing efficient .NET applications.

Chapter 3, Memory Allocation and Data Structures, dives into the critical concepts of memory partitioning and allocation within the .NET framework. Understanding how memory is divided and allocated is essential for optimizing application performance and preventing memory-related issues.

Chapter 4, Memory Leaks and Resource Management, tackles one of software development's most common and challenging issues: memory leaks. Understanding and preventing memory leaks is essential for maintaining the performance and stability of .NET applications. This chapter begins by defining what memory leaks are and how they occur in managed environments despite the presence of automatic garbage collection.

Chapter 5, Advanced Memory Management Techniques, explores sophisticated strategies and techniques for optimizing memory management in .NET applications. Building on the foundational concepts in earlier chapters, this section focuses on high-level methods to enhance application performance, reduce memory consumption, and ensure efficient resource utilization.

Chapter 6, Memory Profiling and Optimization, introduces techniques and tools essential for profiling and optimizing memory usage in .NET applications. The chapter explains the importance of memory profiling and how it helps identify inefficiencies and potential memory leaks that can degrade application performance.

Chapter 7, Low-Level Programming, explores the intricate world of low-level programming within the .NET, offering a deeper understanding of memory management from a closer perspective. We explore how advanced developers can leverage low-level programming techniques to optimize memory usage and enhance application performance.

Chapter 8, Performance Considerations and Best Practices, provides actionable insights, real-world examples, and practical guidelines to help developers build high-performing applications tailored to each specific environment. By the end of this chapter, readers will be equipped with a robust set of best practices and performance optimization techniques that can be applied to desktop, web, and cloud-based .NET solutions, ensuring their applications run smoothly and efficiently in any context.

Chapter 9, Final Thoughts, reflects on the journey through the intricate world of .NET memory management. We revisit the key concepts and techniques discussed throughout the book, reinforcing the importance of effective memory management in developing high-performance, reliable .NET applications.

To get the most out of this book

To get the most out of this book, readers should ideally be familiar with .NET development and its core components is essential. Since C# is the primary language used for examples and exercises in this book, proficiency in C# programming is essential. A solid grasp of general programming concepts such as data structures (arrays, lists, dictionaries), control structures (loops, conditionals), and basic algorithms is necessary to understand and apply the memory management techniques discussed. Familiarity with development environments like Visual Studio, along with experience in debugging and profiling .NET applications, will enable readers to follow along with practical examples and exercises.

Software/Hardware covered in the book	OS Requirements
C# 11/12	Windows, Mac OS X, and Linux (Any)
Visual Studio 2022 (and upwards)	Windows
Visual Studio Code	Windows, Mac OS X, and Linux (Any)

If you are using the digital version of this book, we advise you to type the code yourself or access the code via the GitHub repository (link available in the next section). Doing so will help you avoid any potential errors related to the copying and pasting of code.

Download the example code files

You can download the example code files for this book from GitHub at `https://github.com/PacktPublishing/Effective-.NET-Memory-Management`. In case there's an update to the code, it will be updated on the existing GitHub repository.

We also have other code bundles from our rich catalog of books and videos available at `https://github.com/PacktPublishing/`. Check them out!

Conventions used

There are a number of text conventions used throughout this book.

`Code in text`: Indicates code words in text, database table names, folder names, filenames, file extensions, pathnames, dummy URLs, user input, and Twitter handles. Here is an example: "The developer called the `GC.Collect()` function, which forces a collection event."

A block of code is set as follows:

```
struct Point
{
    public int X;
    public int Y;
}
```

Bold: Indicates a new term, an important word, or words that you see onscreen. For example, words in menus or dialog boxes appear in the text like this. Here is an example: "A variable is declared to have a **type** and a **name**."

> **Tips or important notes**
> Appear like this.

Get in touch

Feedback from our readers is always welcome.

General feedback: If you have questions about any aspect of this book, mention the book title in the subject of your message and email us at `customercare@packtpub.com`.

Errata: Although we have taken every care to ensure the accuracy of our content, mistakes do happen. If you have found a mistake in this book, we would be grateful if you would report this to us. Please visit `www.packtpub.com/support/errata`, selecting your book, clicking on the Errata Submission Form link, and entering the details.

Piracy: If you come across any illegal copies of our works in any form on the Internet, we would be grateful if you would provide us with the location address or website name. Please contact us at copyright@packtpub.com with a link to the material.

If you are interested in becoming an author: If there is a topic that you have expertise in and you are interested in either writing or contributing to a book, please visit authors.packtpub.com.

Share Your Thoughts

Once you've read *Effective .NET Memory Management*, we'd love to hear your thoughts! Scan the QR code below to go straight to the Amazon review page for this book and share your feedback.

https://packt.link/r/1835461042

Your review is important to us and the tech community and will help us make sure we're delivering excellent quality content.

Download a free PDF copy of this book

Thanks for purchasing this book!

Do you like to read on the go but are unable to carry your print books everywhere?

Is your eBook purchase not compatible with the device of your choice?

Don't worry, now with every Packt book you get a DRM-free PDF version of that book at no cost.

Read anywhere, any place, on any device. Search, copy, and paste code from your favorite technical books directly into your application.

The perks don't stop there, you can get exclusive access to discounts, newsletters, and great free content in your inbox daily

Follow these simple steps to get the benefits:

1. Scan the QR code or visit the link below

https://packt.link/free-ebook/978-1-83546-104-4

2. Submit your proof of purchase

3. That's it! We'll send your free PDF and other benefits to your email directly

1

Memory Management Fundamentals

Memory management refers to controlling and coordinating a computer's memory. Using proper memory management techniques, we can ensure that memory blocks are appropriately allocated across different processes and applications running in the **operating system (OS)**.

An OS facilitates the interaction between applications and a computer's hardware, enabling software applications to interface with a computer's hardware and overseeing the management of a system's hardware and software resources.

OSs orchestrate how memory is allocated across several processes and how space is moved between the main memory and the device's disk during executions. The memory comprises blocks that are tracked during usage and freed after processes complete their operation.

While you may not need to understand all the inner workings of an OS and how it interacts with applications and hardware, it is essential to know how to write applications that make the best use of the facilities that OSs make available to us, so that we can author efficient applications.

In this chapter, we will explore the inner concepts of memory management and begin to explore, at a high level, the following topics:

- The fundamentals of memory management
- How garbage collection works
- The pros and cons of memory management
- The effects of memory management on application performance

By the end of this chapter, you should better appreciate the thought process that goes into ensuring that applications make the best use of memory, and you will understand the moving parts of memory allocation and deallocation.

Let's begin with an overview of how memory management works.

Overview of memory management

Modern computers are designed to store and retrieve data during application runtimes. Every modern device or computer is designed to allow one or more applications to run while reading and writing supporting data. Data can be stored either long-term or short-term. For long-term storage, we use storage media such as hard disks. This is what we call **non-volatile** storage. If the device loses power, the data remains and can be reused later, but this type of storage is not optimized for high-speed situations.

Outside of data needed for extended periods, applications must also store data between processes. Data is constantly being written, read, and removed as an application performs various operations. This data type is best stored in **volatile** storage or memory caches and arrays. In this situation, the data is lost when the device loses power, but data is read and written at a very high speed while in use.

One practical example where it's better to use volatile memory instead of non-volatile memory for performance reasons is cache memory in computer systems. Cache memory is a small, fast type of volatile memory that stores frequently accessed data and instructions to speed up processing. It's typically faster than accessing data from non-volatile memory, such as **hard disk drives** (**HDDs**) or **solid-state drives** (**SSDs**). When a processor needs to access data, it first checks the cache memory. If the data is in the cache (cache hit), the processor can retrieve it quickly, resulting in faster overall performance. However, if the data is not in the cache (cache miss), the processor needs to access it from slower, non-volatile memory, causing a delay in processing.

In this scenario, using volatile memory (cache memory) instead of non-volatile memory (HDDs or SSDs) improves performance because volatile memory offers much faster access times. This is especially critical in systems where latency is a concern, such as high-performance computing, gaming, and real-time processing applications. Overall, leveraging volatile memory helps reduce the time it takes to access frequently used data, enhancing system performance and responsiveness.

Memory is a general name given to an array of rapidly available information shared by the CPU and connected devices. Programs and information use up this memory while the processor carries out operations. The processor moves instructions and information in and out of the processor very quickly. This is why caches (at the CPU level) and **Random Access Memory** (**RAM**) are the storage locations immediately used during processes. CPUs have three cache levels: L1, L2, and L3. L1 is the fastest, but it has low storage capacity, and with each higher level, there is more space at the expense of speed. They are closest to the CPU for temporary storage and high-speed access. *Figure 1.1* depicts the layout of the different cache levels.

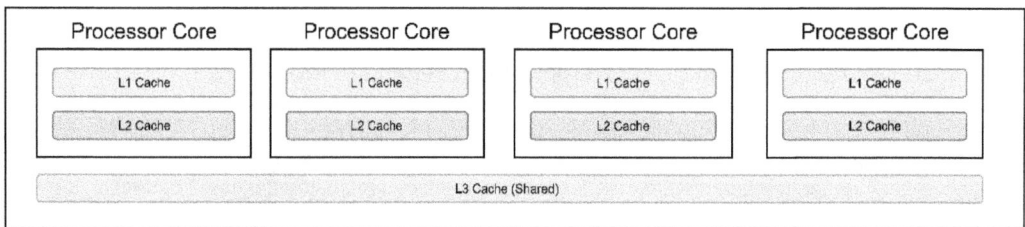

Figure 1.1 – The different cache levels in a CPU and per CPU core

Each processor core contains space for L1 and L2 caches, which allow each core to complete small tasks as quickly as possible. For less frequent tasks that might be shared across cores, the L3 cache is used and is shared between the cores of the CPU.

Memory management, in principle, is a broad-brush expression that refers to techniques used to optimize a CPU's efficient usage of memory during its processes. Consider that every device has *specifications* that outline the CPU speed and, more relevantly, storage and memory size. The amount of memory available will directly impact the types of applications that can be supported by the device and how efficiently the applications will perform on such a device. Memory management techniques exist to assist us, as developers, in ensuring that our application makes the best use of available resources and that the device will run as smoothly as possible while using the set amount of memory to handle multiple operations and processes.

Memory management considers several factors, such as the following:

- **The capacity of the memory device**: The less memory that is generally available, the more efficient memory allocation is needed. In any case, efficiency is the most critical management factor, and failing here will lead to sluggish device performance.

- **Recovering memory space as needed**: Releasing memory after processes run is essential. Some processes are long-running and require memory for extended periods, but shorter-running ones use memory temporarily. Data will linger if memory is not reclaimed, creating unusable memory blocks for future processes.

- **Extending memory space through virtual memory**: Devices generally have volatile (memory) and non-volatile (hard disk) storage. When volatile memory is at risk of maxing out, the hard disk facilitates additional memory blocks. This additional memory block is called a **swap** or **page file**. When the physical memory becomes full, the OS can transfer less frequently used data from RAM to the swap file to free up space for more actively used data.

Devices with CPUs generally have an OS that coordinates resource distribution during processes. This resource distribution is relative to the available hardware, and the performance of the applications is relative to how efficiently memory is managed for them.

A typical has three major areas where memory management is paramount to the performance of said device. These three areas are hardware, application, and the OS. Let's review some of the nuances that govern memory management at these levels.

Levels of memory management

There are three levels where memory management is implemented, and they are as follows:

- **Hardware**: RAM and **central processing unit** (**CPU**) caches are the hardware components that are primarily involved in memory management activities. RAM is physical memory. This component is independent of the CPU, and while it is used for temporary storage, it boasts much more capacity than the caches and is considerably slower. It generally stores larger and less temporary data, like data needed by long-running applications.

- The **memory management unit** (**MMU**), a specialized hardware component that tracks logical or virtual memory and physical address mappings, manages the usage of both RAM and CPU caches. Its primary functions include allocation, deallocation, and memory space protection for various processes running concurrently on the system. *Figure 1.2* shows a high-level representation of the MMU and how it maps to the CPU and memory in the system.

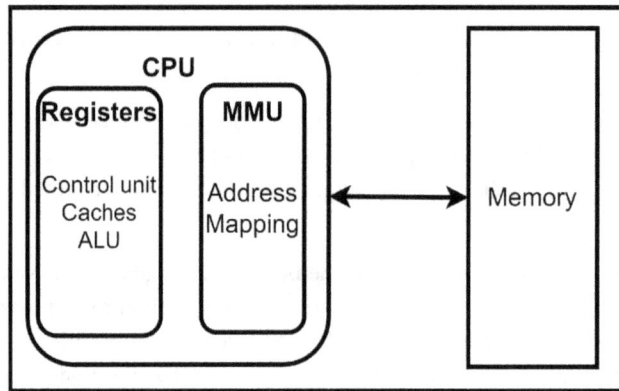

Figure 1.2 – The CPU and MMU and how they connect to memory spaces in RAM

- **OS**: Systems with CPUs tend to have a system that orchestrates processes and operations at a root level. This is called an **operating system** (**OS**). An OS ensures that processes are started and stopped successfully and orchestrates how memory is distributed across the different memory stores for many processes. It also tracks the memory blocks to ensure it knows which resources are being used and by what process to reclaim memory as needed for the following process. OSs also employ several memory allocation methods to ensure the system runs optimally. If physical memory runs out, the OS will use virtual memory, a pre-allocated space on the device's storage medium. Storage space is considerably slower than RAM, but it helps the OS to ensure that the system resources are used as much as possible to prevent system crashes.

- **Application**: Different applications have their memory requirements. Developers can write memory allocation logic to ensure that the application controls memory allocation and not the other systems, such as the MMU or the OS. Two methods are generally used by applications to manage their memory allocation are:

  - **Allocation**: Memory is allocated to program components upon request and is locked exclusively for use by that component until it is no longer needed. A developer can manually and explicitly program allocation logic or automatically allow the memory manager to handle allocation using the allocator component. The memory manager option is usually included in a programming language/runtime. If it isn't, then manual allocation is required.

  - **Recycling**: This is handled through a process called garbage collection. We will look at this concept in more detail later, but in a nutshell, this process will reclaim previously allocated and no-longer-in-use memory blocks. This is essential to ensure that memory is either in use or waiting to be used, but not lost in between. Some languages and runtimes automate this process; otherwise, a developer must provide logic manually.

Application memory managers have several hurdles to contend with. It must be considered that memory management requires the CPU, which will cause competition for the system resources between this and other processes. Also, each time memory allocation or reclamation happens, there is a pause in the application as focus is given to that operation. The faster this can happen, the less obvious it is to the end user, so it must be handled as efficiently as possible. Finally, the more memory is allocated and used, the more fragmented the memory becomes. Memory spreads across non-contiguous blocks, leading to higher allocation times and slower read speeds during runtime.

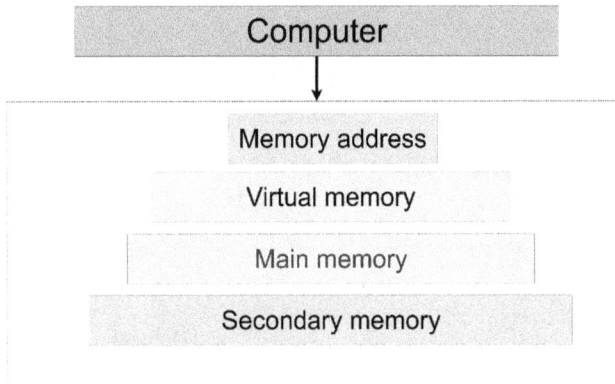

Figure 1.3 – A high-level representation of a computer's memory hierarchy

Each component in a computer's hardware and OS has a critical role to play in how memory is managed for applications. If we are to ensure that our devices perform at their peak, we need to understand how resource allocation occurs and some of the potential dangers that poor resource management and allocation can lead to. We will review these topics next.

Fundamentals of memory management and allocation

Memory management is a sufficiently challenging technique to incorporate in application development. One of the top challenges is knowing when to retain or discard data. While the concept sounds easy, the fact that it is an entire field of study speaks volumes. Ideally, programmers wouldn't need to worry about the details in between, but knowing different techniques and how they can be used to ensure maximum efficiency is essential.

Contiguous allocation is the oldest and most straightforward allocation method. When a process is about to execute, and memory is requested, the required memory is compared to the available memory. If sufficient contiguous memory can be found, then allocation occurs, and the process can execute successfully. If an adequate amount of memory contiguous memory blocks cannot be found, then the process will remain in limbo until sufficient memory can be found.

Figure 1.4 shows how memory blocks are aligned and how the assignment is attempted in contiguous allocation. Conceptually, memory blocks are sequentially laid out, and when allocation is required, it is best to place the data being allocated in blocks beside each other or contiguously. This makes read/write operations in applications that rely on the allocated data more efficient.

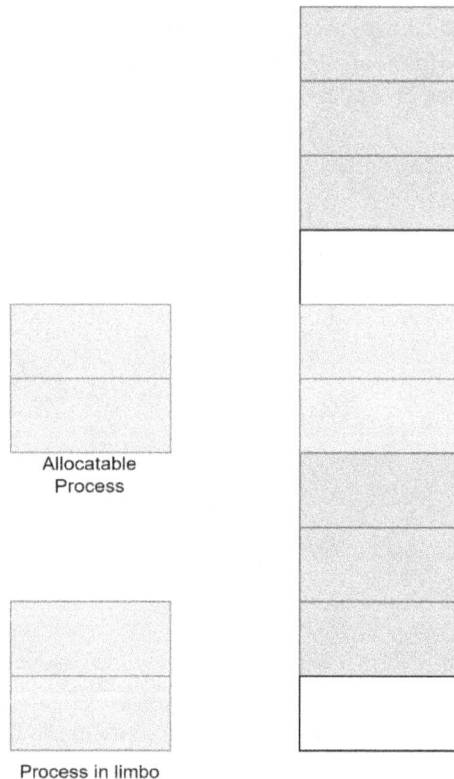

Allocatable
Process

Process in limbo

Figure 1.4 – How contiguous allocation works

The preference for contiguously allocated blocks is evident when we consider that contiguous memory blocks are more accessible to read and manipulate than non-contiguous blocks. One drawback, however, is that memory might not be used effectively since the entire allocation must be successful, or the allocation will fail. For this reason, memory might not get allocated to smaller contiguous blocks.

As developers, we can use the following tips as guidelines to ensure that contiguous allocation occurs in our applications:

- **Static allocation** – We can ensure that we use variables and data structures where a fixed size is known and allocated at application runtime. For instance, arrays are allocated contiguously in memory.

- **Dynamic allocation** – We can manually manage memory blocks of fixed sizes. Some languages, such as C and C++, allow you to allocate memory on the fly using functions such as `malloc()` and `calloc()`. Similarly, you can prevent fragmentation by deallocating memory when it is no longer in use. This ensures that memory is being used and freed as efficiently as possible.

- **Memory pooling** – You can reserve a fixed memory space when the application starts. This fixed space in memory will be used exclusively by the application for any resource requirements during the runtime of the application. The allocation and deallocation of memory blocks will also be handled manually, as seen previously.

These techniques can help developers write applications that ensure contiguous memory allocation as and when necessary for certain systems and performance-critical applications.

With contiguous memory, we have the options of stack and heap allocation. Stack allocation pre-arranges the memory allocation and implements it during compilation, while heap allocation is done during runtime. Stack allocation is more commonly used for contiguous allocation, and this is a perfect match since allocation happens in predetermined blocks. Heap allocation is a bit more difficult since the system must find enough memory, which might not be possible. For this reason, heap allocation suits non-contiguous allocation.

Non-contiguous allocation, in contrast to contiguous allocation, allows memory to be allocated across several memory blocks that might not be beside each other. This means that if two blocks are needed for allocation and are not beside each other, then allocation will still be successful.

Figure 1.5 displays memory blocks to be assigned to a process, but the available slots are at opposite ends of the contiguous block. In this model, the process will still receive its allocation request, and the memory blocks will be used efficiently as the empty spaces are used as needed.

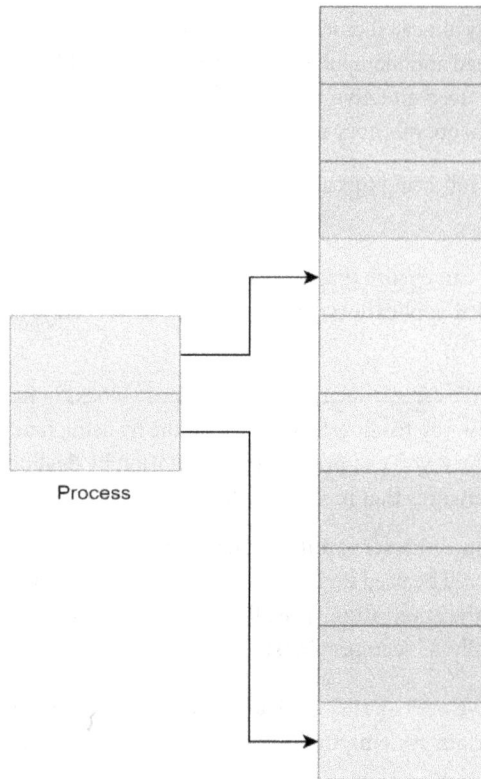

Figure 1.5 – How non-contiguous allocation works, where empty
memory blocks are used even when they are separated

This method, of course, comes at the expense of optimal read/write performance, but it does help an application move forward with its processes since it might not need to wait too long before memory can be found to fulfill its requests. This also leads to a common problem called fragmentation, which we will review later.

Even with the techniques and recommendations, there are many scenarios where a poor implementation of memory management can affect the robustness and speed of programs. Typical problems include the following:

- **Premature frees**: When a program gives up memory but attempts to access it later, causing a crash or unexpected behavior.

- **Dangling pointers**: When a program ends but leaves a dangling reference to the memory block it was allocated.

- **Memory leak**: When a program continually allocates and never releases memory. This will lead to memory exhaustion on the device.

- **Fragmentation**: Fragmentation is when a solid gets split into many pieces. Programs operate best when memory is allocated linearly. When memory is allocated using too many small blocks, it leads to poor and inadequate distribution. Eventually, despite having enough spare memory, it can no longer give out big enough leagues.

- **Poor locality of reference**: Programs operate best when successive memory accesses are nearer to each other. Like the fragmentation problem, if the memory manager places the blocks a program will use far apart, this will cause performance problems.

As we have seen, memory must be handled delicately and has limitations we must be aware of. One of the most significant limitations is the amount of space available to an application. In the next section, we review how memory space is measured.

Units of memory storage

It is essential to know the different units of measurement and overall sizes that specific keywords represent in memory management. This will give us a good foundation for discussing memory and memory usage.

- **Bit**: The smallest unit of information. A bit can have one of two possible numerical values (1 and 0), representing logical values (true and false). Multiple bits combine to form a binary number.

- **Binary number**: A numerical (usually an integer) value formed from a sequence of bits, or ones and zeros. Each bit in the sequence represents a value to the power of 2, with each 1 contributing to the sum of the given value. To convert a binary number to decimal, multiply each digit from left to right by the power of 2. The rightmost digit gets the lowest power.

  For example, the binary number 1101 represents $1 * 8 + 1 * 4 + 0 * 2 + 1 * 1$ to give a total of 13. *Figure 1.6* shows a simple table with the binary positions relative to the power of 2.

Power	2^7	2^6	2^5	2^4	2^3	2^2	2^1	2^0
Base 10 Values	128	64	32	16	8	4	2	1
13					1	1	0	1
37			1	0	0	1	0	1
132	1	0	0	0	0	1	0	0

Figure 1.6 – A simple binary table with example values

- **Binary code**: Binary sequences representing alphanumerical and special characters. Each bit sequence is assigned to specific data. The most popular code is ASCII code, which uses 7-bit binary code to represent text, numbers, and other characters.

- **Byte**: A byte is a sequence of 8 bits that encodes a single character using specified binary code. Since bit and byte begin with the letter b, an uppercase B is used to depict this data size. It also serves as the base unit of measurement, where increments are usually in the thousands (Kilo = 1000, Mega = 1,000,000, etc.)

Now that we understand memory management and its importance, let's look closely at memory and how it works to ensure that our applications run smoothly.

The fundamentals of how memory works

When considering how applications work and how memory is used and allocated, it is good to have at least a high-level understanding of how computers see memory, the states that memory can exist in, and how algorithms decide how to allocate it.

For starters, each process has its own virtual address space but will share the same physical memory. When developing applications, you will work only with the virtual address space, and the garbage collector allocates and frees virtual memory for you on the managed heap. At the OS level, you can use native functions to interact with the virtual address space to allocate and free virtual memory for you on native heaps.

Virtual memory can be in one of three states:

- **Reserved**: The memory block is available for your use and can't be accessed until it's committed
- **Free**: The memory block has no references and is available for allocation
- **Committed**: The block of memory is assigned to physical storage

Memory can become fragmented as memory gets allocated and more processes are spooled up. This means that, as mentioned earlier, the memory is split across several memory blocks that are not contiguous. This leads to *holes* in the address space. The more fragmented memory becomes, the more difficult it becomes for the virtual memory manager to find a single free block large enough to satisfy the allocation request. Even if you need a space of a specific size and have that amount of space available cumulatively, the allocation attempt might fail if it cannot happen over a single address block. Generally, you will run into a memory exception (like an OutOfMemoryException in C#) if there isn't enough virtual address space to reserve or physical space to commit. See *Figure 1.5* for a visual example of how fragmentation might look. The process that has been allocated memory has to check in two non-contiguous slots for relevant information. There is a free space in memory during the process runtime, but it cannot be used until another process requests it. This is an example of fragmented memory.

We need to be careful when allocating memory in terms of ordering the blocks to be allocated relative to each new object or process. This can be a tedious task, but thankfully, the .NET runtime provides mechanisms to handle this for us. Let's review how .NET handles memory allocation for us.

Automatic memory allocation in .NET

Each OS boasts unique and contextually efficient memory allocation techniques. The OS ultimately governs how memory and other resources are allocated to each new process to ensure efficient resource utilization relative to the hardware and available resources.

When writing applications using the .NET runtime, we rely on its ability to allocate resources automatically. Because .NET allows you to write in several languages (C#, C++, Python, etc.), it provides a **common language runtime** (**CLR**) that compiles the original language(s) into a single runtime language called **managed code**, which is executed in a **manager execution environment**.

With managed code, we benefit from cross-language integration and enhanced security, versioning, and deployment support. We can, for example, write a class in one language and then use a different language to derive a native class from the original class. You can also pass objects of that class between the languages. This is possible given that the runtime defines rules for creating, using, persisting, and binding different **reference types**.

The CLR gives us the benefit of automatic memory management during managed execution. You do not need to write code to perform memory management tasks as a developer. This eliminates most, if not all, of the negative allocation scenarios that we previously explored. The runtime reserves a contiguous region of address space for each initialized new process. This reserved address space is the **managed heap**, which is initially set to the base address of the managed heap.

All reference types, as defined by the CLR, are allocated on the managed heap. When the application creates its first reference type instance, memory is allocated at the managed heap's base address. For every initialized object, memory is allocated in the contiguous memory space following the previously allocated space. This allocation method will continue with each new object, if address space is available.

This process is faster than unmanaged memory allocation since the runtime handles the memory allocation through pointers, which makes it almost as fast as allocating memory directly from the CPU's stack. Because new objects allocated consecutively are stored contiguously in the managed heap, an application can access the objects quickly.

The allocated memory space must continuously be reclaimed to ensure an effective and efficient operation. You can rely on a built-in mechanism called a **garbage collector** to orchestrate this process, and we will discuss this at a high level next.

The role of the garbage collector

Garbage collection is the process that governs how programs release memory space that is no longer being used for their operations. This process serves as an automatic memory manager by managing the allocation and release of memory for an application.

Programming languages that support automatic garbage collection free developers from the need to write specific code to perform memory management tasks. Languages that implement automatic memory management allow us to build applications without accounting for common problems such as memory leaks or an application attempting to access freed memory for an already freed object.

Each language handles garbage collection differently, and it is crucial to appreciate how it works in your context. As mentioned, the CLR in .NET implements it automatically, but additional libraries may be required in low-level programming languages such as C. For instance, C developers must handle allocation and deallocation using the `malloc()` and `dealloc()` functions. In contrast, it is not recommended for a C# developer to handle this as it is already taken care of.

Recall that in C#, allocation happens through a managed heap, and objects are placed in contiguous spaces in memory. In contrast, in C, objects are placed where there is free memory, and locations are tracked through a linked list. Memory allocation will work faster in CLR-supported languages since the allocation is done linearly, ensuring a contiguous allocation process. In C, memory must be traversed to find the next available slot, adding additional time to the allocation process. We will review the details of the allocation process of the CLR in the next chapter.

Here are some additional benefits of the garbage collector:

- Allocates objects on the managed heap efficiently

- Reclaims memory from objects no longer being used so that memory is available for future allocations

- Provides memory safety by ensuring an object can't claim memory allocated for another object

The garbage collector boasts an optimized engine that performs collection operations at the best possible time based on static fields, local variables on a thread's stack, CPU registers, GC handles, and the finalized queue from the application's roots. Each root should refer to an object on the managed heap or have a null value. The garbage collector can ask the rest of the runtime for these roots and will use this list to create a graph containing all the objects accessible from the roots. Any unreachable object is classified as garbage, and the memory that it is using is released.

Garbage collection happens under one of these situations:

- The operating system or host has notified that there is low memory.

- Memory being used by the allocated objects on the managed heap exceeds an acceptable threshold.

- The developer called the `GC.Collect()` function, which forces a collection event. This is not generally required since the GC operates automatically.

The managed heap the GC uses to manage allocation is divided into three sections called **generations**. Let's take a closer look at how these generations work and the pros and cons of this mechanism.

Garbage collection in .NET

The GC in .NET has three generations labeled 0, 1, and 2. Each generation is dedicated to tracking objects based on their expected lifetime. Generation 0 stores short-lived objects, ranging to Generation 2 for more long-term objects.

- **Generation 0**: This generation stores short-lived objects such as temporary variables. When this generation is full and new objects are to be created, the GC will free up space by examining the objects in generation 0 rather than all objects in the managed heap.

- **Generation 1**: This generation sits between generations 0 and 2. After a GC event in generation 0, objects are compacted and promoted to this generation, where they will enjoy a longer lifetime. When a GC operation is run on this generation, objects that survive get promoted to Generation 2.

- **Generation 2**: Long-lived objects such as static data and singleton objects are stored in this generation. Anything that survives a collection event on this level stays until it becomes unreachable in a future collection. Collections at this level are also called **full garbage collections** since they reclaim all generations in the heap.

The garbage collector has an additional heap for large objects, called the **Large Object Heap** (**LOH**). This heap is used for objects that are 85,000 bytes or more. Collection events on the LOH and Generation 2 generally take a long time, given the size and lifetime of the cleaned objects.

Garbage collection starts with a marking phase, where it finds and creates a list of all currently allocated objects. It then enters a relocating phase, where references related to the surviving objects are updated. Then, there is a compacting phase where space is reclaimed from dead objects, and the surviving objects are compacted. Compaction is simply the process of moving memory blocks beside each other, which, as mentioned before, is a significant factor in the CLR's efficient memory allocation method.

Applications consist of several processes and processes run on threads. A thread is a basic to which the OS allocates processor time. The .NET runtime and CLR manage threads, and when a garbage collection operation begins, all managed threads are suspended except for the thread that triggered the collection event.

It is generally ill-advised to run the `GC.Collect()` method manually for several reasons. This method will pause your application and allow the collector to run. This may cause your application to become unresponsive and degrade its performance. In addition, the process is not guaranteed to free all unused objects from memory, and those still in use by your application will not be collected. This method should only be used when the application no longer uses any objects that the collector previously collected.

The drawback of garbage collection lies in its effect on performance. Garbage collection must periodically traverse the program, inspecting object references and reclaiming memory. This process consumes system resources and frequently necessitates program pauses.

It is easy to see why garbage collection is a fantastic tool that spares us from carrying out manual memory management and space reclamation. Now, let's review some of memory management's impacts on overall application performance.

Impact of memory management on performance

We have seen the importance of proper and efficient memory management in our applications. Fortunately, .NET makes it easier for us developers by implementing automatic garbage collection to clean up objects in between processes.

Memory management operations can significantly affect your application's performance as allocation and deallocation activities require system resources and might compete with other processes in progress. Take, for example, the garbage collection process, which pauses threads while it traverses the different generations to collect and dispose of old objects.

Now, let's itemize and review some of our application's benefits and the potential pitfalls of memory management:

- **Responsiveness**: Efficient memory management can significantly improve the responsiveness of your application. Your program can run smoothly without unexpected slowdowns or pauses when memory is allocated and deallocated judiciously.

- **Speed**: Memory access times are critical for application speed. Well-organized memory management can lead to more cache-friendly data structures and fewer cache misses, resulting in faster execution times.

- **Stability**: Memory leaks and memory corruption are common issues in applications with suboptimal memory management. Memory leaks occur when memory is allocated but never released, leading to a gradual consumption of resources and potential crashes.

- **Scalability**: Applications that manage memory efficiently are more scalable. They can handle large datasets and user loads without running into memory exhaustion issues.

- **Resource Utilization**: Efficient memory management minimizes memory wastage, allowing your application to run on systems with lower hardware specifications. This can widen your application's potential user base and reduce infrastructure costs.

We can expect these tangible benefits when an application appropriately manages memory. Similarly, there can be some adverse effects when the correct measures are not taken.

Impacts of poor memory management

Memory management can significantly negatively impact application performance if not handled properly. Here are some ways in which poor memory management can adversely affect your application's performance:

- **Memory leaks**: Memory leaks occur when an application fails to release any longer-needed memory. Over time, these leaked memory blocks accumulate, consuming more and more memory resources. This can lead to excessive memory usage, reduced available system memory, and, eventually, application crashes or slowdowns.

- **Inefficient memory usage**: Inefficient memory allocation and deallocation strategies can lead to higher memory consumption than necessary. This can result in your application using more memory than it needs, which can slow down the entire system and reduce the responsiveness of your application.

- **Fragmentation**: Memory fragmentation occurs when memory is allocated and deallocated in a way that leaves small, unusable gaps of memory scattered throughout the heap. This fragmentation can lead to inefficient memory usage, challenging allocating contiguous memory blocks for more extensive data structures or objects. This can cause slower memory access times and reduced application performance.

- **Cache thrashing**: Cache memory is much faster to access than main system memory. Poor memory management can lead to the CPU cache frequently being invalidated and reloaded with data from the main memory due to inefficient memory access patterns. This can result in significant performance degradation.

- **Increased paging and swapping**: When an application consumes too much memory, the OS may resort to paging or swapping data between RAM and disk storage. This involves reading and writing data to and from slower disk storage, which can lead to a noticeable slowdown in application performance.

- **Concurrency issues**: In multi-threaded applications, improper memory management can lead to race conditions, data corruption, and other concurrency issues. Conflicting memory accesses by multiple threads can result in unexpected behavior and performance bottlenecks.

- **Increased garbage collection overhead**: In languages with automatic memory management, such as C#, inefficient memory management practices can lead to more frequent garbage collection cycles. These cycles pause the application briefly to clean up unreferenced objects, which can introduce noticeable delays and reduce overall performance.

- **Resource contention**: When an application consumes excessive memory, it can lead to resource contention with other running processes on the same system. This can result in competition for system resources (CPU, memory, I/O), leading to degraded performance for all running applications.

- **Poor scalability**: Applications with inefficient memory management may struggle to scale. As user loads and data sizes increase, the application's memory demands can become a limiting factor, preventing it from handling larger workloads effectively.

When scoping our applications, we must consider the context that the application is for, the device it will run on, and the resources that will be available. This may lead us to choose a particular language or development stack. Let's review some key considerations.

Key considerations

To ensure optimal performance, developers should carefully consider memory management practices and employ appropriate techniques and tools to mitigate these issues.

It is also important to note that one size does not fit all. When considering the memory management technique that will be implemented, as developers, we must consider the following:

- The type of operations being supported and if they will perform optimally based on how memory is allocated and managed

- The target devices and expected system resources, since a mobile device allocation will differ from a computer's allocation

- The target OS, since each will implement its overall management methods

Now that we understand memory management, techniques, and factors, let's review what we have learned.

Summary

This chapter explored the fundamental concepts of memory management in computer systems. Memory management is critical for OSs efficiently utilizing a computer's memory resources.

We discussed memory hierarchy, which consists of various levels of memory, from registers and cache to RAM and storage devices. Understanding this hierarchy is essential for optimizing memory usage. We then reviewed how memory is allocated to different processes or programs running on a computer, allocation strategies, and their implications on system performance.

We also reviewed a popular memory management technique in garbage collection and how it works in .NET. Garbage collection automatically identifies and frees up unused objects or data memory and excuses the developer from writing manual memory management logic. This behavior also reduces memory overhead and improves overall application performance.

This chapter provided a comprehensive overview of memory management in computer systems, emphasizing its importance in ensuring efficient resource utilization, process isolation, and system reliability. It sets the foundation for understanding the intricacies of memory management in modern OSs.

Next, we look at garbage collection in more depth and see how objects are handled.

2

Object Lifetimes and Garbage Collection

Understanding the life cycle of objects is essential for writing efficient, reliable, and maintainable code. As your applications grow in complexity, you need to have an intricate appreciation for how and when objects are created, used, and ultimately released.

As we saw in the previous chapter, the .NET runtime provides a silent and diligent garbage collector, a cornerstone of memory management activities. It is responsible for identifying and reclaiming memory that is no longer in use. In this chapter, we will take a more in-depth look at the garbage collection process, including the various generations, collection modes, and strategies employed by the runtime to optimize performance.

This chapter explores some of the intricacies of memory management in the .NET runtime, and together, we will explore the following:

- Object allocation and deallocation
- Generations and the garbage collection process
- Best practices for managing object lifetimes

Together, we will unravel the mysteries behind how objects are allocated, the factors influencing their lifespan, and the mechanisms in place to ensure timely and efficient memory reclamation. By the end of this chapter, you will have a deeper understanding of garbage collection in .NET and how to use it to your advantage.

Technical requirements

- Visual Studio 2022 (`https://visualstudio.microsoft.com/vs/community/`)

- Visual Studio Code (`https://code.visualstudio.com/`)

- .NET 6/7/8 SDK (`https://dotnet.microsoft.com/en-us/download/visual-studio-sdks`)

Object allocation and deallocation

Let's look more in-depth at how object allocation and deallocation work in programming and what happens behind the scenes. Before we get into the programmatic aspects of allocation and deallocation, we must first understand and appreciate what an object is and why it is an essential construct in programming.

Objects and how they are created

An object is an instantiation of a class in **object-oriented programming** (**OOP**). In every programming language, we construct a variable, which is used to store data between operations in a program. A variable is declared to have a **type** and a **name**. The types that can be used are as follows:

- **Value type**: Uses primitive types such as `bool`, `int`, and `char`. Variables defined by these types hold data in their location.

- **Reference type**: More complex data types, such as classes, arrays, and delegates. These contain pointers to a memory location that holds the data.

Value-type variables are the simplest form of data storage that you can employ in your application. As we have mentioned, they hold the data that they are set to store. In the following code example, you will see how an integer-based variable is defined. When this value is shared with another variable of the same value type, the value is copied directly:

```
int num1 = 3;
int num2 = num1;
num1 = 5;
Console.WriteLine($"num1: {num1}");
Console.WriteLine($"num2: {num2}");
```

The output in *Figure 2.1* shows that the num2 variable took the value directly from num1:

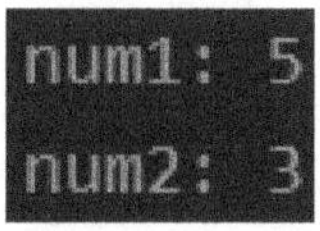

Figure 2.1 – Output of simple value type variables copying values

Reference types allow us to define more complex data types. A commonly used construct for this is a **user-defined class**, which defines variables with more complex data requirements and behaviors than value type variables.

A **class** is a blueprint or template of an entity we wish to represent programmatically. They contain attributes that can represent properties, fields, or member variables. For example, a Person class might have FirstName, LastName, DateOfBirth, and Age attributes. These attributes are defined using existing values or reference types. Access to the attributes can be controlled using access modifiers, which control whether the data can be accessed directly via the defined variable or if they are hidden and only accessed via **methods**.

Methods represent actions the object can perform. For instance, we might want to calculate the person's age based on their DateOfBirth. This calculation can be considered as an action being performed because a method must be called to retrieve the Age value.

All user-defined classes contain a default method called a **constructor**. A constructor is a method with no return type and the same name as the defined class. A developer might define a constructor to perform custom actions, such as assigning initial values to properties.

Here is an example of a Person class:

```
public class Person
{
    // Constructor method
    public Person(string firstName, string lastName, DateOnly
    dateOfBirth)
    {
        FirstName = firstName;
        LastName = lastName;
        DateOfBirth = dateOfBirth;
        _age = DateTime.Now.Year - dateOfBirth.Year;
    }

    // Attributes - Properties and fields
    public string FirstName { get; set; } // property - public access
    public string LastName { get; set; }
    public DateOnly DateOfBirth { get; set; }
    private int _age; // field - not publicly accessible

    // Methods
    public double GetAge()
    {
        return _age;
    }
}
```

Now that we have defined our blueprint, or programmatic representation, of what a person should look like for our application, we can create variables of this reference type that reference the object address somewhere in memory.

We commonly refer to the variables of reference types as objects. However, these variables only contain address details to an object defined in memory. This is why, in C# at least, when we define an instance of the reference type, we need to use the new keyword. This action calls the default constructor of the Person class, which, as the name suggests, constructs an object of the reference type in memory, and the variable stores the address details. Without this new keyword, the variable has a **null reference**.

In the following code example, we initialize two objects and attempt a similar assignment operation as depicted with value type variables in the previous example:

```
// 1 - create variable with a null reference
Person nobody;

// 2 - create a variable to reference to Person object
Person arthur = new Person("Arthur","Wint",new DateOnly(2001,10,25));

// 3 - create a variable to copy the Person object location
Person james = arthur;

// 4 - Update the name property of the copy of the Person
james.FirstName = "James";

// 5 - Print the name values from the copy object reference.
string jamesFullName = $"{james.FirstName} {james.LastName}" ;
Console.WriteLine($"James Full name: {jamesFullName}");

// 6 - Print the name values from the original object reference.
string arthurFullName = $"{arthur.FirstName} {arthur.LastName}" ;
Console.WriteLine($"Arthur Full name: {arthurFullName}");
```

Observe that the output, as shown in *Figure 2.2*, shows that the name value for the arthur object was changed, alongside the targeted james object:

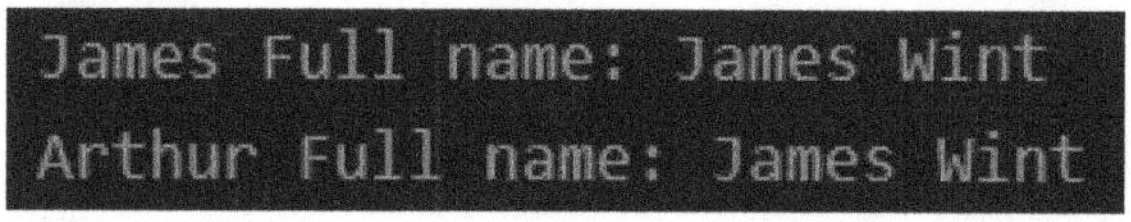

```
James Full name: James Wint
Arthur Full name: James Wint
```

Figure 2.2 – The outcome of reference type assignment and value modification

Once again, these reference type variables store references to the object in memory. So, the following things happened in these statements:

- The `arthur` variable is assigned an address to a new `Person` object in memory
- The `james` variable is assigned the object address from `arthur`
- Since `arthur` and `james` reference the same object in memory, the values are accessible from either variable

This difference in behavior between value type and reference type objects is essential to appreciate as it affects how these objects are created and stored in memory during the allocation process. This is when we begin the conversation about stack and heap allocations, which we will dive into next.

Stack and heap allocation

Stack and **heap** are core components of **memory** in any device. In the context of .NET programming and the **Common Language Runtime** (**CLR**), they are used to allocate objects.

The stack operates as a straightforward **Last-In-First-Out** (**LIFO**) structure. Variables assigned to the stack are directly stored in memory, ensuring rapid access. Stack allocation occurs during program compilation. When a method is called, the CLR marks the stack's top. As the process is executed, data is sequentially pushed onto the stack. Upon completion, the CLR restores the stack to its initial bookmark, effortlessly removing all memory allocations made during the method's execution in a single operation.

The stack generally imposes a size limit; in C#, that default stack size is *1 MB* per thread. Recall that a variable is essentially a space in memory and different data types require different amounts of space. This 1 MB allocation is available for as many variables as possible until garbage collection or cleanups occur. The stack is full when no new data can be inserted in it and if anything else is to be inserted, the system will throw a **stack overflow error**.

The stack aims to complete two operations:

- Track each method that is called and the sequence in which they are called
- Track each variable that is instantiated in each method's scope
- Remove each variable's reference once it is no longer needed after the method has finished executing and the scope no longer exists

In contrast, the heap is a considerably less organized storage space for objects as it allows objects to be allocated, deallocated, and accessed in a random order. Variables assigned to the heap undergo memory allocation during runtime, resulting in comparatively slower access to the allocated memory.

Unlike the stack, the heap does not impose a fixed size limit and can dynamically adjust based on the available virtual memory. Elements stored in the heap are independent, allowing random and untethered access. This allows flexible **allocation** and **deallocation** of memory blocks, empowering developers to allocate a block when needed and release it when no longer necessary.

However, one drawback of the heap's dynamic nature is that it introduces increased complexity in managing memory allocation and deallocation. More careful tracking is required to determine which portions of the heap are currently in use and which are available for allocation.

Now the question is, how do we know what variables get stored in the stack and which ones get stored in the heap? Recall that variables in C# must be declared to a specific data type. A variable is simply an association between a name and a block of memory that lives in the stack or heap. Value type variables are allocated to the stack, and reference type variables are allocated to the heap.

Practically speaking, all variables are stored in the stack, but the reference type variables are only pointers to the value object stored in the heap. *Figure 2.3* shows a depiction of this:

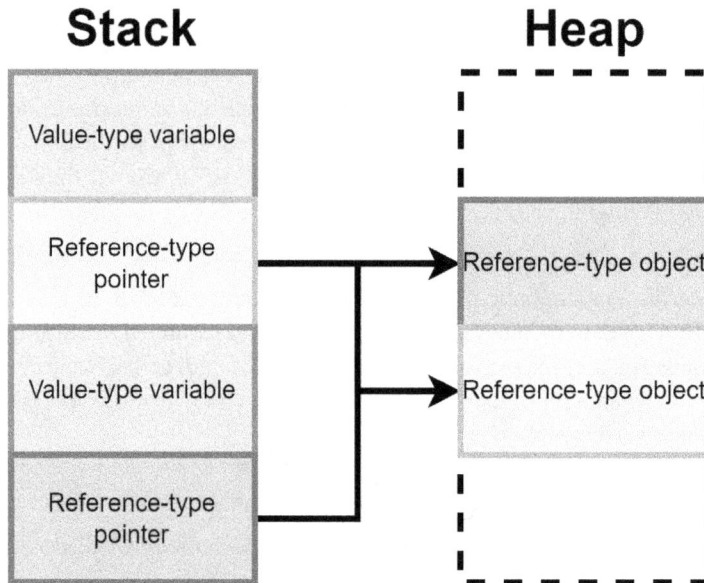

Figure 2.3 – What stacks and heaps store for variable values

One significant point to note regarding stack and heap allocation is that variables are assigned based on the context or scope in which they are defined. For example, a struct is a value type, so a struct defined in a function will be allocated to the stack as expected when that function is called. If the struct is defined in a class, it will be included with the reference type object in the heap.

In the previous section, we reviewed creating and using two int variables, num1 and num2. Because int is a value type, both variables would be allocated on the stack. These variables are considered local variables as they would be executed in the main function of a C# program. We will get one memory block per variable, and because they are defined in the main function, they are also contained in the stack frame for that function.

Figure 2.4 depicts what happened in memory during this code execution:

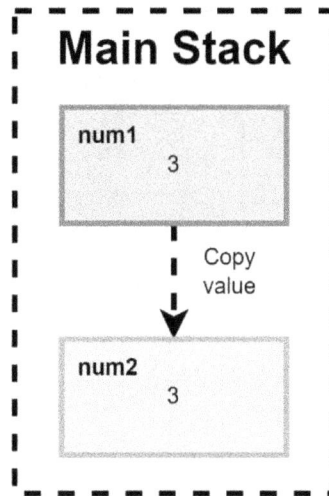

Figure 2.4 – How variables and values are assigned and copied on the stack

We then looked at using a user-defined class called Person to create variables that contain a reference to the same Person object in the heap. Also, notice that we have an object called nobody, which is not initialized. A variable gets created on the stack without a heap object pointer.

In **action 2**, we initialized the variable called arthur using the new keyword, which calls the class's default constructor. Since the constructor is a method, a new stack frame will be created where the variables for that method will now exist. This means that the values passed from the main method into the constructor's parameters are copied from the main stack into the constructor's stack. Finally, the constructor will assign these values to the corresponding properties of the referenced heap object.

Essentially, stacks are created within the executing scope of a method. This can be conceptualized as the code block between an open and a closed curly brace. All variables defined inside those curly braces will exist in a stack created for that code block. Once the close curly brace is reached or that function has finished executing, the stack frame and the variables are removed.

The amount of memory a heap object requires is relative to the amount of space required by the member variables and objects, along with the space needed for the object's header information. This header contains valuable information for internal operations such as garbage collection.

After initializing `arthur`, we created a new variable called `james` in **action 3** and assigned it the same value as `arthur`. In doing so, contrary to what we observed with `num1` and `num2`, the object value itself was not assigned; instead, the address information was. This meant that changing values of the member variables in `james` resulted in the same object being manipulated under a new variable name.

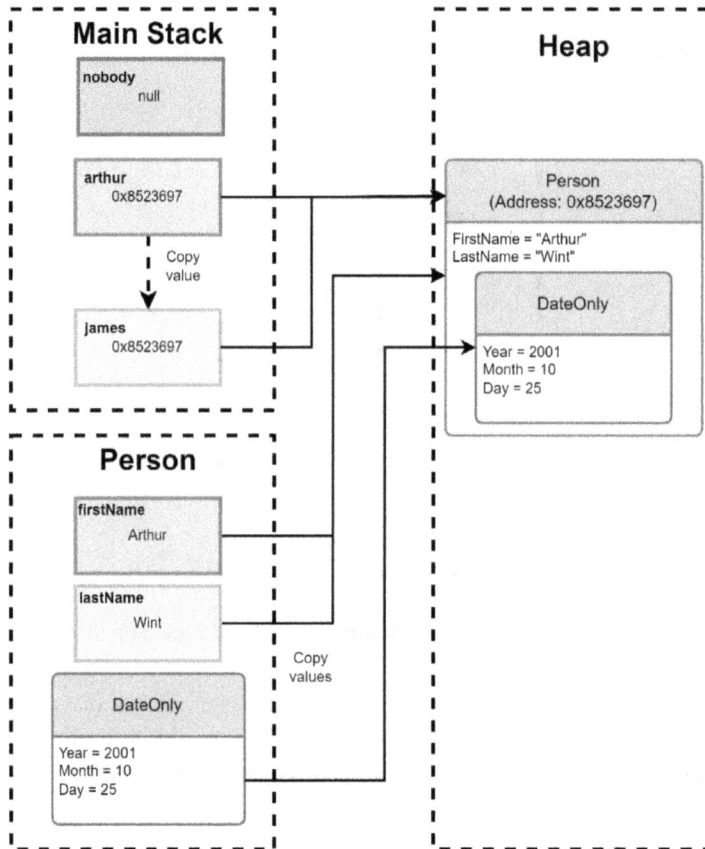

Figure 2.5 – Variables, objects, and values are assigned and copied on the stack and heap

Notice that the `DateOfBirth` property maps to an object in the heap. This property is of the struct type `DateOnly`, which would be allocated on the stack. It, however, is a part of the `Person` class, so it will exist on the heap along with the other instantiated properties of the `Person` object. Conceptually, a string is reference-type, so those properties should also be depicted as objects on the heap.

All these allocations happen without us needing to write manual code, thanks to the CLR and **garbage collector (GC)** provided by .NET. With these mechanisms running in the background, the heap and stack are managed, and memory is reclaimed automatically when variables and objects have outlived their purpose. Let's take a more in-depth look at the garbage collector, managed heap, and background operations that help to keep our program's memory usage efficient.

Generations and the garbage collection process

Since the CLR handles the automatic creation and allocation of objects in memory, its most trusted companion is the automatic GC. The GC monitors objects in memory and decides whether they should stay longer based on the program's needs or if they should be destroyed from memory to be reclaimed.

This automatic memory management mechanism eliminates common developer problems, such as the following:

- Forgetting to free an object, which can lead to a memory leak and poor use of system resources
- Attempting to access an object that has already been removed from memory

The .NET runtime has a built-in garbage collection engine orchestrated by the CLR. The CLR employs a garbage collection strategy to ensure that available memory is only occupied by objects still useful for the current processes. This mechanism is called the GC.

When a new process is initialized, the CLR reserves a contiguous space in memory called the **managed heap**. This heap contains pointer information that determines where the next object is to be created in the heap.

The managed heap can be separated into the **small object heap** (**SOH**) and the **large object heap** (**LOH**). Objects greater than or equal to **85,000 bytes** in size are allocated on the LOH. The default threshold value can be configured at the application level for performance tuning. While this threshold can be increased, attempting to store extra-large objects is ill-advised.

The GC ensures that objects are managed in the managed heap automatically. This spares us, the developers, from needing to write code to manage this all the time. The GC monitors objects that have become "garbage" and deletes them during collection cycles. We can also set an object to `null` in code to let the GC know that it can relinquish the space in the managed heap and use it for another object.

An object becomes "garbage" when one or more of the following is true:

- The object instance is unreachable
- There are no references to the object
- The object ceases to exist inside the scope of the code being executed
- The `GC.Collect()` method has been invoked

The GC is constantly monitoring situations where any one of these conditions holds. It will then clean the candidate objects from the managed heap. *Figure 2.6* shows a representation of memory after an application initializes, and the CLR reserves space in memory for the managed heap. You will see where there are:

- **Application roots**: Application roots are objects that contain static fields, local variables, and other general information about the objects the application uses.
- **Reachable objects**: Objects in the heap that are allocated and still reachable.

- **Free space**: Reserved and unallocated space in memory.

- **Unreachable objects**: An object that is considered unreachable. This object meets at least one of the previously mentioned criteria.

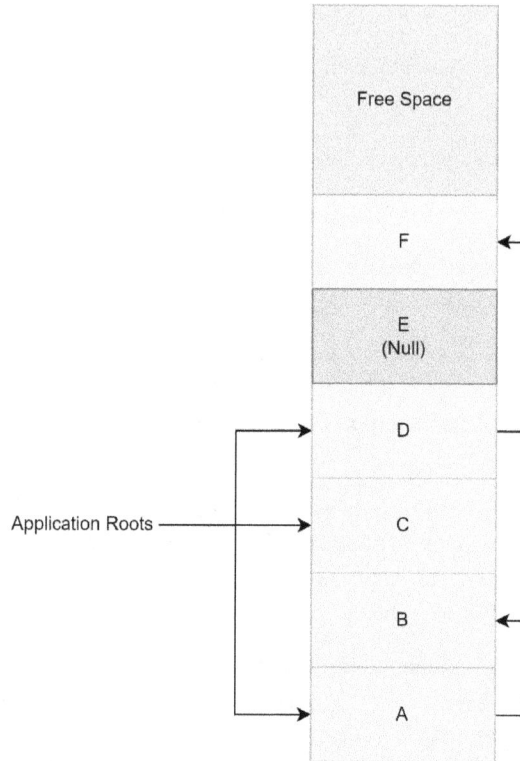

Figure 2.6 – The SOH before garbage collection

The GC has three phases when it assesses objects in the managed heap:

- **Mark**: The GC determines which objects are still reachable or not set to null references.

- **Relocate**: The GC updates the references to the objects that will be compacted.

- **Compact**: **Compacting** is the act of moving objects that have been deemed reachable towards the end of the segment. The GC will recover the memory occupied by the unreachable objects. This rearrangement of objects ensures that the reserved memory space remains contiguous. You can see an example of memory after compacting in *Figure 2.7*:

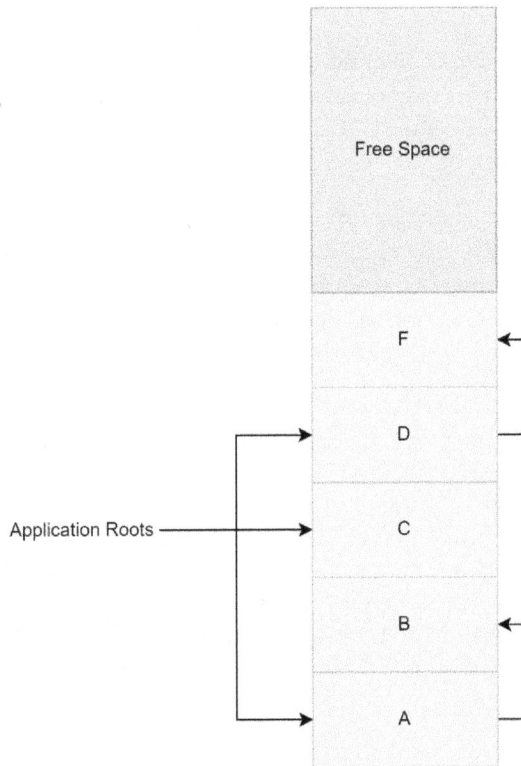

Figure 2.7 – The SOH after garbage collection or compacting

The GC categorizes objects into three groups or generations, tracking which objects should be removed now, relocated later, or left alone for a while. Three generations are associated with objects on the SOH, 0, 1, and 2.

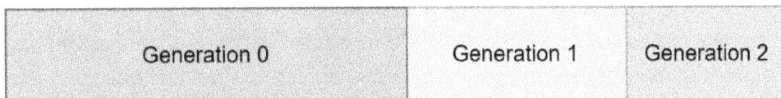

Figure 2.8 – The different generations on the SOH

Generation 0 is the youngest tier in the memory management hierarchy, housing short-lived objects, such as temporary variables. GC operations occur frequently in this generation, and newly allocated objects automatically become part of generation 0. However, if these objects are sizable, they are placed on the LOH. You may consider the LOH as Generation 3, a physical category, but it is logically collected alongside Generation 2.

Most objects are reclaimed through garbage collection in generation 0 and do not persist to the subsequent generation. When generation 0 reaches full capacity, triggering an attempt to create a new object, the GC initiates a collection to free up address space. This collection focuses on examining objects in generation 0, and often, the reclamation of memory in this generation alone is sufficient to allow the application to continue creating new objects.

Generation 1 serves as a buffer between short-lived and long-lived objects. Following a garbage collection of Generation 0, the surviving objects are promoted to Generation 1. This act reduces the frequency of scrutiny that longer-lived objects will go through. If a Generation 0 collection fails to recover enough memory to create a new object, the GC can perform sequential collections of Generation 1 and Generation 2. Surviving objects in Generation 1 are promoted to Generation 2.

Generation 2 stores long-lived objects, such as those in server applications with static data persisting for the entire process. Objects surviving a collection in generation 2 remain there until determined to be unreachable in a future collection. Objects on the LOH are also collected during Generation 2 collections.

Do note that the GC interacts with the LOH differently since compacting and moving large objects around requires a lot of system resources. Large objects are not allocated within the same address space as small objects but are assigned a distinct location within the process's address space. Let's explore this next.

Garbage collection in the LOH

As previously mentioned, a large object is defined as any object with a size of 85,000 bytes or greater. The LOH is not compacted, as copying large objects comes at the expense of application performance. Instead, it deallocates the memory taken up by substantial objects in the LOH once they are no longer in use. Consequently, address space fragmentation can occur among large objects in the process, potentially triggering an OutOfMemoryException. *Figure 2.9* shows what the LOH looks like at the time of allocation and after some deallocation:

LOH w/ allocations

Allocated
Allocated
Allocated
Allocatied

LOH fragmented

Free Space
Allocated
Free Space
Allocatied

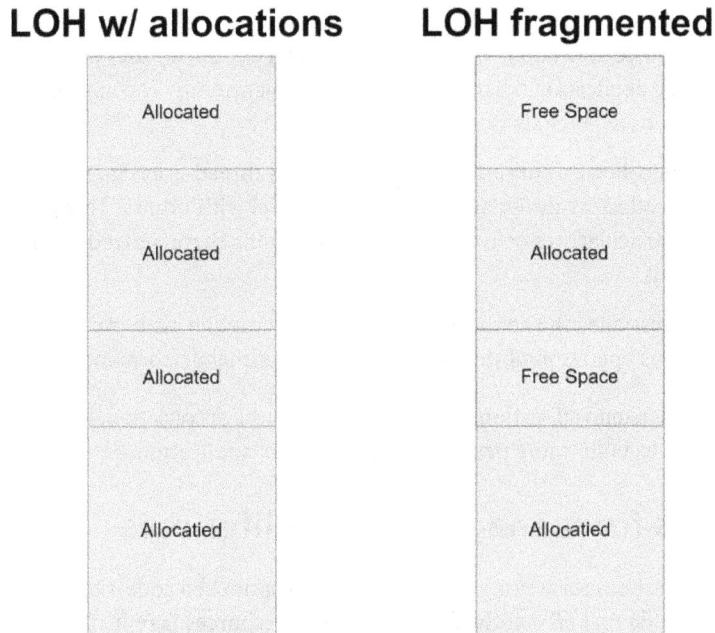

Figure 2.9 – The LOH before and after deallocation and possible fragmentation

Large objects are immediately categorized as part of generation 2, skipping generations 0 and 1 entirely. Therefore, allocating large objects only for resources intended for prolonged retention is advisable. Allocating short-lived large objects may result in more frequent collection cycles in generation 2, negatively impacting performance.

Typically, large objects can be large strings (such as XML or JSON) or large byte arrays. Large objects are transparent to you, and it is not evident that they are being treated differently in the background unless you run into performance issues or other unexplainable behaviors in your application that will require you to dig deeper.

.NET supports two garbage collection modes: **Workstation GC** and **Server GC**. The choice between these modes depends on the specific application or workload, necessitating the implementation of different modes to optimize garbage collection:

- Workstation GC tailors the GC for client-side applications, optimizing it for low-latency garbage collections to minimize thread suspension time and enhance the user experience. The GC assumes that other applications are concurrently running on the machine, preventing excessive consumption of CPU resources.

- Server GC customizes the GC for server-side applications, focusing on maximizing throughput and efficient resource utilization. The GC assumes that it exclusively uses the machine's resources and that no other application (client or server) runs concurrently. The managed heap is divided into multiple sections, each dedicated to a CPU.

- During garbage collection initiation, one specialized thread is assigned per CPU, allowing each thread to collect its designated section in parallel with others. This parallel collection mechanism is particularly effective for server applications characterized by uniform behavior in worker threads.

- This feature necessitates the application to run on a computer with multiple CPUs to ensure true simultaneous operation of threads and achieve optimal performance enhancements.

Now that we better understand allocation, deallocation, and garbage collection, let's review some good practices that will help us build more performant and efficient applications.

Best practices for managing object lifetimes

Let's look at some suggestions for writing more efficient and optimized code. One of the fundamental principles of writing code that efficiently uses your system resources is reducing object lifetimes. If we author code with this in mind, we can maximize resource utilization and guarantee smoother and more efficient applications.

We have just reviewed how the GC determines which objects should be cleaned from memory and which should stay for longer. We need to write our code to ensure that objects created for a purpose get marked for deletion once that purpose is fulfilled. As an object's lifespan increases, its memory consumption grows over time. This increased consumption can give rise to issues such as memory leaks or unwarranted strain on the GC, leading to a decline in application performance. By proficiently handling object lifetimes and ensuring they persist only as long as essential, we can enhance memory efficiency and boost the application's overall performance.

Several techniques can be used to ensure that objects get cleaned up on time and only live for as long as they are needed. We have seen the conditions under which the GC will determine that objects need to be cleaned, and, in some cases, this will happen organically because of how the code is written. In other cases, we need to be more deliberate and aware of the possible implications of objects and write defensive code that ensures that our objects do not get overlooked by the GC when appropriate.

The first technique that we will review is to use local variables.

Using local variables

Using local variables in C# can benefit the GC because local variables have a limited scope and lifetime that's tied to the block of code where they are declared. When a method or block of code finishes execution, the local variables within that scope go out of scope, and the garbage collector can reclaim their memory.

Since local variables are automatically cleaned up by the runtime when they go out of scope, you don't have to manage the memory deallocation explicitly. Another benefit is that they typically require memory for a shorter duration. This reduces the overall memory pressure on the system because memory is released more promptly, and the GC has less work to do. With less work, of course, comes quicker and easier operations. The GC will more quickly identify and collect short-lived objects during regular collection cycles, leading to more efficient and faster garbage collection.

Local variables are also less prone to causing memory leaks because their lifespan is well-defined within a specific scope.

The following code snippet shows how a method can be authored to use an in-scope variable. Open and closed curly braces generally define the scope and where a block of code starts and finishes its execution. Variables defined within that scope only exist when that block of code is being executed:

```
public void ExampleMethod()
{
    // Object with limited scope
    var localVar = new SomeObject();

    // Use localVar within this method only
    // ...

} // localVar goes out of scope and is eligible for garbage collection
```

So, to reiterate, it is best practice to ensure that variables only exist within the scope that they are needed.

The next management method that we will review is using statements.

Using statements

In C#, the using statement is a convenient and concise way to ensure the proper disposal of objects that implement the IDisposable interface. We primarily use this statement to manage variables and objects in a precise manner when completing an operation. It ensures that the objects defined in the scope are automatically disposed of by calling the Dispose method on the object when the code block is exited. This will happen whether the block is exited normally or due to an exception.

The IDisposable interface provides a standard mechanism for releasing resources from unmanaged resources, such as file handles, network connections, or database connections. Objects that embody such functionality may go untracked by the GC under normal circumstances. We will explore this further in upcoming chapters, but know that C# provides types that embody the functionality required for these operations and implement the IDisposable interface. This allows us to release these resources explicitly when they are no longer needed.

An example of an **unmanaged type** that implements the IDisposable interface is StreamReader. This type is typically used to interact with text files. If misused, the Dispose method of the class will not be called and, while the managed memory associated with the StreamReader will be reclaimed by the GC eventually, the unmanaged resources (such as file handles) might not be released promptly. This is a recipe for potential issues such as file locks and decreased system resources. An example of this misuse is shown here:

```
public class FileReader
{
    public string ReadFileContent(string filePath)
    {
        // StreamReader is not used within a using block
        StreamReader streamReader = new StreamReader(filePath);

        // Reading file content
        string content = streamReader.ReadToEnd();

        // StreamReader.Dispose() is not called explicitly

        return content;
    }
}
```

Now, even though the method's scope is done, the unmanaged resources are not reclaimed, and a GC operation is not carried out to release the resources being used.

The using statement allows us to define a scope specifically for the object in use so that as soon as it finishes its operation, the remaining lines of code may be executed, and all the resources will be freed:

```
public class FileReader
{
    public string ReadFileContent(string filePath)
    {
        string content = string.Empty;
        // StreamReader used within a using block
        using (StreamReader streamReader = new StreamReader(filePath))
        {
            // Reading file content
            content = streamReader.ReadToEnd();

            // StreamReader.Dispose() is automatically called
        }
        // The rest of the method code executes
        // ...
```

```
        return content;
    }
}
```

In this code example, we properly dispose of the `StreamReader` object before the rest of the method scope is completed.

We saw earlier in this chapter that we have reference and value types. Each one presents its own benefits and challenges. Let's review how choosing the correct one can help with our application.

Use optimized data types and data structures

Using optimized data structures in C# involves designing your data structures and managing references to minimize memory usage, reduce unnecessary allocations, and ensure efficient garbage collection.

One step in the right direction is to use value types as much as possible, which is practical. Value types are more efficient than reference types in specific scenarios due to their characteristics and how they are stored in memory. Value types are stored directly in the memory location where they are declared, which can reduce heap allocations.

The efficiency of value types versus reference types depends on the specific use case and the nature of the data being manipulated. Both have pros and cons, and the choice between them should be based on the requirements of the application and the characteristics of the data being handled. In the following code example, we use a `struct` (value type) instead of a class (reference type) to define a type called `Point`:

```
struct Point
{
    public int X;
    public int Y;
}
```

Another method to reduce object creation overhead is to be strategic in our code. Ensure that you are wisely creating and initializing your objects. Avoid creating an object with no reference or repeatedly creating objects in loops, or frequently called methods. It is better to define one object and then use it in the loop, changing the value of each iteration or passing it as a parameter to a method for manipulation:

```
// Instead of creating a new StringBuilder in each iteration
StringBuilder sb = new StringBuilder();
for (int i = 0; i < 1000; i++)
{
    sb.Append(i);
}
```

Use weak reference types for short-lived data

We can also use **weak reference types** to ensure our objects are short-lived. In C#, a weak reference type allows the GC to collect an object even if there are references to it. Unlike a strong reference, a weak reference does not prevent the referenced object from being eligible for garbage collection.

This can be beneficial in scenarios where you want to hold a reference to an object without preventing it from being reclaimed by the GC when memory usage is high. We can define and use a weak reference object as follows:

```
WeakReference<Person> weakReference = new WeakReference< Person>(new
Person());
if (weakReference.TryGetTarget(out Person obj))
{
    // The object is still alive
    // Use 'obj'
}
else
{
    // The object has been collected
}
```

The `TryGetTarget` method determines if the object is still alive. If the object has been collected, the method returns `false`, and the reference is considered stale and ready for collection. Weak references are helpful in scenarios where strong references might lead to memory leaks but should be used carefully, as improper use might lead to unexpected behavior. The object referenced by a weak reference may be collected at any time, so the reference should always be checked before attempting to use the object.

You can create either a short, weak reference or a long weak reference. A short, weak reference's target becomes null once the associated object undergoes garbage collection. Like any other managed object, the weak reference itself is subject to garbage collection. The parameterless constructor for weak reference is employed to create a short, weak reference.

A long weak reference, however, persists even after the object's `Finalize` method has been invoked. This allows the object to be reconstructed; however, the object's state remains unpredictable. Set the Boolean parameter to `true` in the `WeakReference` constructor to establish a long weak reference, as seen in the following code snippet:

```
WeakReference<Person> weakReference = new WeakReference< Person>(new
Person(), true);
```

It's worth noting that if the object's type lacks a `Finalize` method, the functionality aligns with that of a brief weak reference. In such cases, the weak reference is valid only until the target is collected, a process that can occur any time after the finalizer is executed.

We will discuss finalization and the Dispose pattern in *Chapter 4*.

In summary, weak references in C# are a powerful tool for managing object lifetimes, preventing memory leaks, and improving garbage collection efficiency in scenarios where objects are only needed if they are still in use. They are handy when you want to avoid retaining objects longer than necessary or where strong references might lead to circular references.

Use object pooling

Object pooling in .NET refers to a technique used to manage and reuse objects, typically to improve performance and resource utilization in applications. Instead of creating and destroying objects dynamically as needed, object pooling maintains a pool of pre-initialized objects that can be reused across multiple requests or threads.

A pool comprises pre-initialized objects that can be reserved and released across multiple threads. Pools may specify allocation guidelines, such as constraints, predetermined capacities, or expansion rates.

The .NET Framework provides support for object pooling through the `System.Runtime.Caching.ObjectPool<T>` class, which allows developers to create and manage pools of objects of a specific type T. Additionally, third-party libraries and frameworks, such as the `Microsoft.Extensions.ObjectPool` package in ASP.NET Core, offer more advanced object pooling features and customization options.

Let's review an example of object pooling while using our `Person` class:

```
private static readonly ObjectPool<Person> PersonPool = new
DefaultObjectPool<Person>(new PersonPooledObjectPolicy());
    public static void Main(string[] args)
    {
        // Simulate creating and using Person objects
        for (int i = 0; i < 5; i++)
        {
            UsePerson();
        }
    }
    private static void UsePerson()
    {
        // Get a Person instance from the pool
        var person = PersonPool.Get();

        try
        {
            // Use the Person instance
            Console.WriteLine($"Using Person: {person.Name},
            Age: {person.Age}");
```

```
        }
        finally
        {
            // Return the Person instance to the pool
            PersonPool.Return(person);
        }
    }
}
public class PersonPooledObjectPolicy : IPooledObjectPolicy<Person>
{
    public Person Create()
    {
        // Simulate expensive object creation
        Console.WriteLine("Creating new Person object");
        return new Person();
    }

    public bool Return(Person person)
    {
        // Reset the state of the Person object before returning it to
        // the pool
        person.Name = null;
        person.Age = 0;
        return true;
    }
}
```

We create a static ObjectPool<Person> called PersonPool, which uses a DefaultObjectPool implementation and is initialized with a PersonPooledObjectPolicy. Inside the UsePerson method, we obtain a Person instance from the pool using PersonPool.Get(). We use the acquired Person instance and then return it to the pool using PersonPool.Return(person). The PersonPooledObjectPolicy class implements the IPooledObjectPolicy<Person> interface. It provides methods to create and return Person instances to and from the pool. In this example, we simulate expensive object creation in the Create method by writing a message to the console. In the Return method, we reset the state of the Person object before returning it to the pool.

While this is a simple example, you might find more tangible benefits from this suggestion when managing database connections or HTTP client objects. In these cases, the pool of connection or client objects will live and get reused for as long as the application runs, reducing the overhead of opening and closing connections between operations.

Remember that optimizing data structures for each object's lifetime often involves a trade-off between memory usage and performance. It's essential to profile your application and identify specific areas where optimization can provide tangible benefits based on the application's requirements. Now, let's review what we have learned in this chapter.

Summary

In this chapter, we took a more intimate look at garbage collection, object allocation and deallocation, and some techniques for ensuring that we write efficient applications. Understanding object and reference lifespans in our system can help us, as developers, to make good decisions when authoring code.

We always want to ensure that, while the .NET runtime gives us a wonderful background tool to automatically clean up our mess at times, we do not create unnecessary mess as well.

The CLR manages how objects are allocated in memory and reserves a contiguous block for the application's needs. In this contiguous block, objects are created and cleaned up based on the needs of the process in effect. The significant advantage of having a contiguous block reserved is that there is little to no fragmentation. This makes it easy to find references to objects when needed. This contiguous space is also called the **managed heap**, and the GC constantly polls this heap to find objects that are no longer required.

The GC categorizes objects based on their immediate usage and will instantly remove short-lived objects when they are no longer needed. These objects reside in a GC category called Generation 0. Objects with longer lifespans are promoted to Generations 1 and 2, where the GC will check on them less frequently.

Large objects are allocated on a separate heap where they are not affected by the GC operations as much. Large objects require more resources, so their management must be done differently. In a later chapter, we will dive into LOH management in more detail.

Finally, we reviewed some best practices for managing objects and their lifetimes. Once again, we must make good decisions during development. Something as simple as defining a local variable instead of a global one, using a value type instead of a reference type, or defining a specific scope for an object that gets created at the expense of locking resources can be the difference between an efficient long-running application, or one that consumes too much memory over time.

In the next chapter, we will take a deeper look at memory allocation and what that looks like for different data types while we continue to explore efficient development techniques.

3

Memory Allocation and Data Structures

We continue our journey into the fundamental aspects that form the bedrock of efficient and robust software systems. Memory management is the silent architect behind every line of code, influencing performance, scalability, and responsiveness. Understanding how memory is allocated and managed is crucial for crafting high-performance applications.

This chapter will guide you through the intricacies of memory allocation and the strategic use of available data structures. Whether you're a seasoned developer or a relatively new developer, grasping these foundations is crucial to optimizing your code.

This chapter explores the following:

- Memory allocation mechanisms
- Choosing the right data structures
- Handling large objects and arrays

Overall, this understanding contributes to writing more efficient and robust code and enables developers to design and implement maintainable, scalable, and optimal systems in terms of both speed and resource usage. Applications that efficiently handle memory and data structures are more scalable. Understanding how large objects and arrays are handled helps manage memory more effectively, particularly in high-load and large-scale applications. This knowledge helps prevent memory fragmentation and make informed decisions about data structure design, which can avert performance degradation over time.

The importance of choosing the most appropriate data structures to develop efficient, scalable, and performant applications cannot be overstated. Data structures serve as the building blocks of software applications, providing efficient means of organizing and accessing data. In .NET development, choosing the proper data structure can significantly impact an application's performance and scalability. Choosing between arrays, lists, dictionaries, or more specialized structures such as trees and graphs and understanding their characteristics and suitability for specific tasks is imperative for crafting high-performance solutions.

This chapter seeks to unlock the secrets of memory management and data structures, empowering you to create software that runs efficiently and scales gracefully. Let us begin with a deep dive into memory allocation mechanisms built into the .NET runtime.

Technical requirements

We will be comparing performance metrics of various data types. As a result, we will need to ensure a suitable development environment. We'll need the following to write our code:

- Visual Studio 2022 (https://visualstudio.microsoft.com/vs/community/)

- Visual Studio Code (https://code.visualstudio.com/).

- .NET 6/7/8 SDK (https://dotnet.microsoft.com/en-us/download/visual-studio-sdks)

Memory allocation mechanisms

At the core of every application's performance lies the intricacies of memory allocation. As discussed in the previous chapter, the .NET runtime uses a built-in stack and heap to store and track variables and objects created during an application's runtime. Understanding these mechanisms equips you with the knowledge to make informed decisions and write code optimizing memory resources. .NET employs a managed memory model through mechanisms designed to optimize memory usage and mitigate common memory-related issues such as memory leaks and buffer overflows.

Before exploring more .NET allocation mechanisms, let's examine other popular languages, such as C/C++ and Java, and see how these techniques differ from those in .NET.

Allocation mechanisms in C/C++

Each language has its own methods of handling allocations. C is considered a foundational language as many other languages (such as C++, Java, and C#) have taken their cues from its syntax and how it performs. Each derived language, however, attempts to improve the developer experience and, to some extent, the efficiencies.

C can be considered foundational for the following reasons:

- **Proximity to hardware**: C is a low-level language that provides direct access to hardware resources, making it suitable for system programming, embedded systems, and device drivers. Its syntax and features align closely with the underlying hardware architecture, allowing developers to write efficient, portable code.

- **Portability**: C programs can be easily ported across different platforms and architectures with minimal modifications. The availability of C compilers for a wide range of systems facilitates this portability, making it a popular choice for developing cross-platform software.

- **Efficiency**: C is known for its runtime performance and memory usage efficiency. It allows developers to control memory allocation and management directly, enabling optimizations tailored to specific hardware constraints and performance requirements.

- **Flexibility**: C balances high-level and low-level programming paradigms, allowing developers to work at various levels of abstraction. Its rich set of features, including pointers, structures, and memory management primitives, provides flexibility in implementing complex algorithms and data structures.

Regarding memory management, we have already established that C requires the developer to manually control allocation and deallocation throughout the application code. This is the bedrock of why the preceding points are possible. C gives developers direct control over memory allocation and deallocation, allowing them to manage memory resources according to specific application requirements.

This control enables fine-tuning of memory usage for optimal performance and resource utilization. Developers can allocate memory only when needed and deallocate it as soon as it's no longer required, minimizing memory overhead and fragmentation. This makes memory usage patterns more transparent and easier to understand. This predictability simplifies memory optimization and debugging processes.

Appreciating manual memory management in C lays a solid foundation for understanding memory management in other languages and environments. Without getting into the details of how the code would look in C or C++, we can explore the concepts at a high level:

- The functions necessary for allocation are found in the `<stdio.h>` header files.

- `malloc()` can allocate memory for a data structure dynamically.

- Before attempting to access the variable pointing to the newly allocated memory, we need to check whether the allocation was successful by checking whether the pointer is NULL. This step is crucial since the allocation can fail due to insufficient memory. This risk presents itself if there isn't enough contiguous free memory to satisfy the allocation request. This can occur if memory is allocated and deallocated in a **non-uniform pattern**, leading to fragmented memory segments. In this case, `malloc()` will return NULL.

Systems with memory limits or quotas will limit the amount of memory available for allocation. This may happen in resource-constrained environments or when running applications with restricted memory access. Hardware failures, operating system errors, or interference from other software components can also cause allocation operations to fail unexpectedly.

Once we have verified that the allocation was successful, we can manipulate the array based on the operation's requirements. In this snippet, we add and then print some values. Finally, memory allocated during the program's execution must be freed before the program closes. We release the allocated memory using the `free()` function to prevent leaks.

C++ was developed as an extension to C, where C++ retains most of the control structures, functions, standard libraries, and memory management methods. Some of the significant improvements that C++ introduced are as follows:

- **Object-Oriented Programming (OOP)**: C++ supports OOP paradigms, including classes, inheritance, polymorphism, and encapsulation. These features allow for the creation of modular, reusable, and maintainable code, making C++ suitable for large-scale software development.

- **Standard Template Library (STL)**: It provides a collection of generic data structures (e.g., vectors, lists, and maps) and algorithms (e.g., sorting and searching) that are highly optimized and easy to use. It also significantly reduces the amount of manual coding required for everyday tasks, improving productivity and code quality.

- **Templates**: Allow developers to write generic code that works with any data type. They enable the creation of flexible and efficient algorithms and data structures that can adapt to different data types without sacrificing performance.

- **Exception handling**: This provides a structured mechanism for handling runtime errors and abnormal conditions. Exceptions allow for graceful error recovery and propagation, improving the robustness and reliability of C++ programs.

- **Standardization**: C++ is standardized by the ISO C++ standard, which ensures language consistency and compatibility across different compilers and platforms. The standardization process helps maintain C++'s stability and facilitates interoperability between different C++ code bases.

C and C++ offer manual memory management, meaning developers are responsible for explicitly allocating and deallocating memory. Because C++ introduces OOP features, memory management in C++ often involves managing objects created using these OOP features, which is not present in C. In C++, objects can be allocated to the heap using the new operator, which creates objects with a variable size or lifetime. Memory manually allocated on the heap must also be explicitly deallocated to prevent memory leaks using the `delete` operator. Pointers are similarly used to store memory addresses.

C++ provides constructors and destructors, which are special member functions called automatically when objects are created and destroyed, respectively. Constructors initialize object state, including memory allocation, while destructors clean up resources, including deallocating memory. C does not have such language features.

C++ also introduces smart pointers (`std::unique_ptr`, `std::shared_ptr`, and `std::weak_ptr`) as part of its standard library. Smart pointers automatically manage the memory of dynamically allocated objects, ensuring proper deallocation when they go out of scope. They provide safer and more robust memory management than raw pointers used in C.

Other upgraded memory management features include tools such as **Resource Acquisition Initialization** (**RAII**), which leverages the deterministic destruction behavior of objects to manage resources automatically. The core idea behind RAII is to tie the lifespan of a resource (such as dynamically allocated memory, file handles, mutexes, or network connections) to the lifespan of an object representing that resource.

Like a C program, we include the header files necessary for input/output and memory operations, `<iostream>` and `<memory>`, respectively. In the `main` function, we use `new` to allocate memory for an array of size `10` dynamically. This creates a raw pointer, which we must check to ensure the allocation succeeded by checking for a `nullptr` value. After using the array in the operations, we free the memory using the `delete[]` syntax. If memory is dynamically allocated, we must use the `delete` keyword to free up any allocated memory.

Next, we use `std::unique_ptr`, a smart pointer provided by the C++ Standard Library. We dynamically allocate memory for a single integer and use it for some operations. In contrast to raw pointers, smart pointers automatically manage memory deallocation and automatically release the allocation once the function's scope is completed.

While C# is not directly based on C/C++, it is influenced by syntax and includes some foundational features. We can see where C# is regarding memory management techniques and features since it is based on the .NET runtime, which automatically handles memory allocation and deallocation. We discussed this in the previous chapter and reviewed code snippets of how much easier it is to define and use objects without considering cleaning them up afterward.

Java and C# are more similar as they are both based on C/C++ and have their runtimes, which help to manage objects automatically. We will compare their methods next.

Allocation mechanisms in Java

Java was created by Sun Microsystems, Inc., and first released in 1995. The difference between how Java and other programming languages worked was revolutionary. Different languages use a compiler to translate the original syntax into instructions for a specific type of computer. In contrast, the Java compiler instead turns code into something called **bytecode**, which is then interpreted by the **Java Runtime Environment** (**JRE**) or the **Java Virtual Machine** (**JVM**). The JRE interprets the bytecode and translates it for the host computer or device. This allows Java to be used for many platforms and be run virtually anywhere and on any device. This portability led to its popularity in various software development contexts and, especially at the time, on the internet.

Java's design principles emphasize reliability and error-handling mechanisms. Features such as solid type-checking, exception handling, and automatic memory management (garbage collection) contribute to the stability and predictability of Java programs, reducing the likelihood of runtime errors and crashes. It is also a pure OOP language, where everything is treated as an object. It also includes a robust standard library called **Java Standard Edition - Java SE**, which provides a range of classes and APIs for everyday input/output operations, networking, concurrency, and data manipulation tasks.

When comparing Java and .NET, several recurring themes will emerge regarding how memory is allocated and managed during an application's runtime. Both Java and .NET offer automatic memory management through garbage collection.

Java objects are stored in an area known as the **heap**, which is initialized when the JVM starts and can dynamically adjust its size during runtime. As the heap reaches capacity, garbage collection occurs where objects no longer in use are identified and removed, creating room for new objects. The JVM also uses memory beyond just the heap. Components, such as Java methods, thread stacks, and native handles, have their own allocated memory space, distinct from the heap.

The JVM heap is often segmented into two sections: the **nursery** (or young space) and the **old** space. The nursery serves as a dedicated area for allocating new objects. Once it fills up, a **young collection process** takes place, promoting objects that have existed in the nursery for a particular duration to the old generation space and remaining objects that are cleared. When the old generation space reaches capacity, a garbage collection process, called an **old collection**, occurs. Utilizing a nursery is rooted in the observation that many objects are short-lived. So, the young collection process is optimized to identify and relocate newly allocated objects in the nursery swiftly. Typically, this process releases memory more efficiently than an old or garbage collection in a single-generational heap setup, which lacks a nursery.

Several similarities in garbage collection and memory management between Java and .NET stem from their shared goals and underlying principles of managing memory safely and efficiently, especially within managed runtime environments. Both Java and .NET aim to simplify development and improve application performance and reliability.

Here are some reasons for their similarities:

- **Automatic memory management**: Both Java and .NET were designed to abstract away the complexity of manual memory management from the developer. Both frameworks help prevent memory leaks and dangling pointer errors common in languages requiring manual memory management (such as C and C++).

- **Managed runtime environment**: Java runs on JVM, while .NET applications run on the **Common Language Runtime (CLR)**. These environments provide a layer of abstraction over the operating system, allowing applications to be somewhat platform-independent. JVM and CLR handle memory allocation and garbage collection as part of managing program execution, leading to similarities in how memory is managed.

- **Generational garbage collection**: Java and .NET use generational garbage collection algorithms based on the observation that most objects die young. Memory is divided into generations, and objects that survive garbage collection cycles are moved to older generations. This approach optimizes garbage collection by focusing on areas of memory where garbage is most likely to be found, reducing the overall impact on application performance.

- **Mark and sweep algorithm**: The "mark and sweep" algorithm involves marking objects that are still in use and then sweeping away those that are not, freeing up memory for new objects. Java and .NET use this method when deciding what objects should be collected and which survive.

- **Just-in-time compilation**: Both Java and .NET compile code into an intermediate language (Java bytecode for Java, and **Common Intermediate Language** (**CIL**) for .NET) that is then compiled into native code at runtime by a **Just-In-Time** (**JIT**) compiler. While JIT compilation is primarily about performance optimization, it also impacts memory management by optimizing the code's memory usage at runtime.

In summary, Java and .NET share similar garbage collection and memory management objectives, which simplify development, ensure application safety and security, and improve performance within managed runtime environments. These similarities reflect broader trends in software development toward abstraction, automation, and platform independence.

Now that we have seen how other popular languages allocate and handle memory, let's compare .NET allocation performance with theirs.

.NET allocation performance

Memory allocation in .NET is generally considered to be quite efficient, especially in the context of managed languages and environments. Over the years, the .NET runtime and the **garbage collector** (**GC**) have been optimized to perform well for various applications. Furthermore, ongoing improvements in .NET Core enhance the efficiency of garbage collection and memory allocation, reflecting Microsoft's commitment to improving .NET performance.

.NET Core's memory allocation improvements over the .NET Framework result from several architectural changes and optimizations. Since .NET Core was designed from the ground up to be more modular and lightweight than the traditional .NET Framework, memory allocation in .NET Core is generally faster and more efficient than in its predecessor as it boasts faster memory allocation is the result of its modern architecture, an optimized JIT compilation, modular design, and innovative features such as **Span<T>** and **Memory<T>**. These advancements make .NET Core well-suited for building high-performance applications that run efficiently on multiple platforms.

Performance comparisons between .NET, Java, C, and C++ depend heavily on the context, including the type of application, the environment in which it's run, how the code is written, and the specific workload or task being performed. Ultimately, the performance of a .NET application compared to one written in Java, C, or C++ depends on how well the application is designed and implemented.

A well-optimized C or C++ program might outperform .NET or Java for compute-bound tasks. For applications where productivity, safety, and rapid development are priorities and where the performance overhead of runtime checks and garbage collection is acceptable or can be minimized through optimizations, .NET or Java might be the better choice. Performance testing and profiling are essential to deciding which platform or language suits a specific project.

Regardless of the application type or language, it is always best to write your code consciously and be mindful of the pros and cons of the different data types. We will discuss some primary considerations when developing an application and why one data type might better fit specific scenarios.

Choosing the best data structures

Choosing the proper data structure in .NET and C# development is crucial for your application's performance, scalability, maintainability, and complexity. The data structure you choose influences how data is organized in memory, how efficiently you can access and manipulate that data, and how well your application scales with increasing data volumes or user demand.

Using the proper data structure can lead to more understandable and maintainable code. It can make the intention behind your code more straightforward to others (or yourself in the future), reduce the risk of errors, and make the code base easier to extend and refactor. The correct data structure can help ensure that your application scales effectively as the data or number of users grows. Structures that provide efficient operations can help prevent performance bottlenecks.

In this section, we will also write sample code and implement benchmark tests to report statistics showing performance differences. Before we begin our programming adventure, though, we should explore and understand a commonly overlooked concept in development called the **Big O notation** and how it helps us decide which programming construct is best for a scenario. Let us discuss this concept to understand how it helps us choose the most suitable data structures relative to our operations.

Big O notation

This mathematical notation describes the upper bound of an algorithm's complexity, representing the worst-case scenario regarding time or space (memory) as a function of the input size (n). It gives a high-level understanding of how an algorithm's performance scales as data increases without getting bogged down in the hardware details or the programming language used. Big O notation focuses on the two main factors that contribute to the complexity of an algorithm:

- **Time complexity**: How the execution time of an algorithm increases with the size of the input data. Understanding the time complexity of operations in data structures (e.g., accessing an element in an array is $O(1)$, while searching for a component in a linked list is $O(n)$) allows developers to choose the appropriate data structure that meets the performance requirements of their application.

- **Space complexity**: How the amount of memory an algorithm requires increases with the input data size. Space complexity analysis helps understand algorithms' memory usage. Choosing data structures with efficient memory usage is crucial in environments with limited resources or applications dealing with large datasets. It can prevent memory from running out and ensure the application's scalability.

Big O notation provides a formula for analyzing and comparing algorithms' efficiency. This is especially important in manipulating and interacting with collection types. This analysis helps choose the most efficient approach for a given context, significantly impacting performance and scalability. Some standard notation that you will see include the following:

- **O(1) – Constant time**: Execution time or space is fixed and does not change with the input data size.

- **O(log n) – Logarithmic**: Execution time or space grows logarithmically as the input data increases.

- **O(n) – Linear**: Execution time or space increases linearly with the input data size.

- **O(n log n) – Linearithmic**: Execution time or space increases faster than linear but not as quick as polynomial.

- **O(n^2), O(n^3) – Polynomial**: Execution time or space increases astronomically with the input data size. This means that performance significantly degrades as the data grows.

- **O(2^n), O(n!) – Exponential and factorial**: Extremely inefficient, with execution time or space increasing at an explosive rate as the input size grows.

In software development and computer science, problems can often be solved in multiple ways. Understanding the Big O notation helps evaluate different approaches and optimize existing solutions for better performance and lower resource consumption. Choosing the appropriate structure can significantly reduce computational costs and improve performance.

Let us review an example illustrating how Big O notation can help determine the best data structure for this scenario in a **user management system** where we must frequently check whether a user exists in the system, add new users, and remove users. The system should perform these operations as efficiently as possible:

- **Option 1 – List**:

    To find a user in a `List<T>`, you might have to scan each element, which makes the search operation `O(n)`. Adding new users at the end of the list is `O(1)`, but inserting at a specific position can be `O(n)` because it might require shifting elements. Like insertion, deletion can be `O(n)` because it may involve shifting elements after removing an item.

- **Option 2 – Hash set**:

  HashSet<T> is built on a hash table, allowing average-case constant time complexity, O(1), for searches. Inserting a new user is typically O(1) on average, thanks to direct access patterns provided by hashing. Deleting a user is also O(1) on average.

- **Option 3 – Sorted data structure**:

  SortedSet<T> uses a binary search tree under the hood, offering O(log n) search times. Inserting a new user is O(log n) since it needs to maintain the order. Deleting a user is also O(log n) for the same reason as insertion.

If the primary operation is frequent searching, then HashSet<T> is advantageous due to its average-case constant time complexity for search, insert, and delete operations. If the order of elements matters or if there is a need for sorted data, SortedSet<T> might be a better choice despite the logarithmic cost for operations, as it maintains order and provides efficient lookups. If simplicity and ordered iteration over elements are acceptable trade-offs for less efficient insertions and deletions, List<T> might still be suitable.

Now that we appreciate the Big O notation and can use it to help us decide which data type might be best for different scenarios, let's look at the first recommendation for data type selection: we should prefer a struct to a class when possible. Let us now review some steps to set up our code projects, which will be the basis for our benchmarking activities.

Replacing classes with structs wherever possible

The recommendation to replace classes with structs in .NET Core development is not universally applicable. Still, it does come with certain benefits under specific circumstances, primarily due to differences in how the .NET runtime handles value types (such as structs) versus reference types (such as classes).

A class consumes more memory resources than a struct. Unlike a class, a struct doesn't have a **method table** or **object header**. Consider using a struct when replacing a class with only a few data members. A method table, **virtual method table**, or **vtable**, is a mechanism used in OOP languages to support dynamic dispatch, polymorphism, inheritance, and late binding of methods. It is essentially a lookup table associated with each class containing pointers to its methods. When a method is called on an object, the runtime uses the method table to find and invoke the correct method implementation, allowing for behavior such as overriding methods in derived classes.

Method tables introduce overhead for classes since each class has a method table, which consumes memory. This overhead is more pronounced in systems with many classes or when several instances of classes are created. Regarding overall performance, **dynamic dispatch** requires additional runtime checks and indirection to look up the correct method in the method table. This can lead to slightly slower method calls compared to **static dispatch**, where the method to be called is known at compile time.

Structs, being a value type in C# and .NET, do not typically support inheritance or method overriding in the same way that classes do. Since structs are designed to be lightweight and efficient, they do not use method tables for dynamic dispatch. Method calls on structs are resolved at compile time (static dispatch) rather than runtime, eliminating the need for a method table and avoiding its associated overhead.

The object header is a part of the internal representation of objects in many managed runtimes, such as the .NET CLR and the JVM. It contains metadata about the object, including information necessary for the runtime's memory management and synchronization mechanisms. This metadata can include, but is not limited to, a pointer to the object's method table, information for the garbage collector, and synchronization information (used for locking in multithreading scenarios).

Stack-allocated structs have their lifetime managed by the scope in which they are declared, eliminating the need for garbage collection and, by extension, the need for metadata such as that found in the object header. They also cannot inherit from other structs or classes (though they can implement interfaces), so there is no need for the runtime type information that an object header would provide to support dynamic dispatch or polymorphism. While structs can implement interfaces, this is typically handled differently than class inheritance and does not require a method table pointer in every struct instance.

Because of the differences in memory allocation between structs and classes, structs don't have the memory allocation overhead of classes:

- **Structs are value types** typically allocated on the stack, which can be more efficient than heap allocations for classes (reference types). Stack allocations can lead to less pressure on the GC and, potentially, better performance. Since structs are not allocated on the heap (except when boxed or part of a class), they do not incur garbage collection overhead. This can lead to more predictable performance, particularly in high-throughput or low-latency applications.

- **Structs have value semantics**, meaning they are copied on assignment and passed by value to methods. This behavior is different from reference types, which are passed by reference. Value semantics can be beneficial for ensuring immutability and avoiding unintended side effects.

- **Structs are ideal for small, immutable data structures**. The efficiency of stack allocation and the value semantics of structs make them suitable for representing simple values or small data aggregates where immutability is desirable.

- **Structs can offer performance-critical benefits** in sections of an application where minimizing GC overhead is crucial. They can also be implemented as high-frequency operations on small data structures, such as mathematical vectors or coordinates.

- **Structs help to avoid heap allocation overhead**. For scenarios where minimizing heap allocations and GC pressure is essential, using structs can help reduce overhead. This is especially relevant in tight loops or real-time systems where performance consistency is critical.

While there are noticeable performance benefits to be reaped from using structs instead of classes, there are some considerations and limitations:

- While structs avoid GC overhead, they are copied entirely on assignment and when passed to methods. This can result in a performance penalty for large structs compared to reference types, where only the reference is copied.

- When assigned to an object or interface type, structures can be "boxed" into reference types, leading to heap allocation and potential GC pressure. Careful use is necessary to avoid unintentional boxing.

We will review some code examples and benchmark tests to review different scenarios. Since we will write some code, let us create a new folder for this project and navigate to it in your system terminal. Create a new console application using the following dotnet CLI command:

```
dotnet new console
```

Once set, execute the following command to prompt the NuGet package manager to install BenchmarkDotNet:

```
dotnet add package BenchmarkDotNet
```

From now on, we will use this as the basis for some upcoming coding activities. Ensure that you run the test console apps in release mode. Let us discuss some memory allocation concepts first.

Your application's requirements and performance characteristics should guide the choice between structs or classes in .NET Core development. While structs can offer performance benefits in terms of reduced GC pressure and efficient memory allocation, their suitability depends on the size of the data being represented and how it's used within the application. The following code snippet compares benchmarks between structs and classes:

```
public class StructVsClassesBenchmarks
{
    const int items = 100000;

    [GlobalSetup]
    public void GlobalSetup()
    {
        //Write your initialization code here
    }


    [Benchmark]
    public void CreatingInstancesUsingClass()
    {
        MyClass[] myClassArray = new MyClass[items];
        for (int i = 0; i < items; i++)
```

```
        {
            myClassArray[i] = new MyClass();
        }
    }


    [Benchmark]
    public void CreatingInstancesUsingStruct()
    {
        MyStruct[] myStructArray = new MyStruct[items];
        for (int i = 0; i < items; i++)
        {
            myStructArray[i] = new MyStruct();
        }
    }


}
class MyClass
{
    public double Num1 { get; set; }
    public double Num2 { get; set; }
    public double Num3 { get; set; }
}

struct MyStruct
{
    public double X { get; set; }
    public double Y { get; set; }
    public double Z { get; set; }
}
```

This code snippet uses two methods to create two large arrays, one for classes and the other for structs. They are then populated using a for loop. *Figure 3.1* shows the outcome of this benchmark test. In the Program.cs file, write the following to execute the benchmark tests:

```
var summary = BenchmarkRunner.Run<StructVsClassesBenchmarks>();
```

We need to execute this program in release mode since the release configuration builds a version of the app that can be deployed. So, you may use the following command in your terminal:

```
dotnet run -c release
```

Once this command is run, the console window will launch, run the tests, and output information similar to what is displayed in *Figure 3.1*:

```
// * Summary *

BenchmarkDotNet v0.13.12, Windows 11 (10.0.22631.3235/23H2/2023Update/SunValley3)
Intel Core i7-9750H CPU 2.60GHz, 1 CPU, 12 logical and 6 physical cores
.NET SDK 8.0.200
  [Host]     : .NET 8.0.2 (8.0.224.6711), X64 RyuJIT AVX2
  DefaultJob : .NET 8.0.2 (8.0.224.6711), X64 RyuJIT AVX2


| Method                     | Mean       | Error     | StdDev    |
|--------------------------- |-----------:|----------:|----------:|
| CreatingInstancesUsingClass | 4,561.9 us | 144.13 us | 418.15 us |
| CreatingInstancesUsingStruct |   670.5 us |  13.20 us |  26.96 us |

// * Warnings *
MultimodalDistribution
  StructVsClassesBenchamarks.CreatingInstancesUsingClass: Default -> It seems that the distribution can have several modes (mValue = 2
.9)

// * Hints *
Outliers
  StructVsClassesBenchamarks.CreatingInstancesUsingClass: Default  -> 3 outliers were removed (6.31 ms..6.96 ms)
  StructVsClassesBenchamarks.CreatingInstancesUsingStruct: Default -> 3 outliers were removed (747.59 us..755.50 us)

// * Legends *
  Mean   : Arithmetic mean of all measurements
  Error  : Half of 99.9% confidence interval
  StdDev : Standard deviation of all measurements
  1 us   : 1 Microsecond (0.000001 sec)

// ***** BenchmarkRunner: End *****
Run time: 00:01:50 (110.44 sec), executed benchmarks: 2

Global total time: 00:02:13 (133.41 sec), executed benchmarks: 2
// * Artifacts cleanup *
Artifacts cleanup is finished
```

Figure 3.1 – The results of a benchmark comparison between classes and structs

The results show that the struct array assignments occur in a fraction of the time it took for the class array to do the same thing. In practice, we should prefer structs for small, immutable data structures where value semantics are beneficial and use classes for larger, more complex objects or when inheritance is required. As with performance optimization, profiling and testing are crucial to making informed decisions.

Now, let's examine how to reduce allocation overheads and use optimized collection types.

Using optimized collection types

Choosing the right collection type for your needs can significantly impact performance, especially regarding memory usage and execution speed. .NET provides a variety of specialized collection types in the System.Collections, System.Collections.Generic, System.Collections. Concurrent, and System.Collections.Immutable namespaces, among others. Using these optimized collection types effectively requires understanding their characteristics and best use cases.

Here's how to choose and use them:

- **Generic collections (System.Collections.Generic)**: Prefer generic collections such as `List<T>`, `Dictionary<TKey, TValue>`, and `HashSet<T>` over their non-generic counterparts because they offer better type safety, which helps to eliminate runtime errors and the need for costly type checks or casts at runtime, thus improving performance. They are also more performant since their usage avoids boxing and unboxing. Non-generic collections store elements as *objects*, requiring value types to be boxed when added to the collection and unboxed when retrieved. Boxing and unboxing are computationally expensive operations that involve creating a wrapper object on the heap for a value type and then extracting the value type from the wrapper. Generic collections eliminate this overhead by storing value types directly, without boxing, thus improving memory usage and access speed.

- **Concurrent collections (System.Collections.Concurrent)**: Use concurrent collections such as `ConcurrentDictionary<TKey, TValue>`, `ConcurrentQueue<T>`, and `ConcurrentBag<T>` in multithreaded scenarios to ensure thread safety without manually handling synchronization. Many concurrent collections use sophisticated algorithms to minimize locking. For example, `ConcurrentDictionary` uses fine-grained locking or lock-free techniques for read operations, allowing multiple threads to read and write concurrently with less contention. Because concurrent collections handle their synchronization, developers can write more straightforward, concise code. Avoiding manual synchronization reduces the cognitive load and potential for errors, allowing developers to focus on business logic and potentially leading to more efficient and maintainable code. While concurrent collections offer many benefits, they are not a silver bullet for all concurrency problems. We will discuss the details of concurrency and memory management in *Chapter 5*.

- **Immutable collections (System.Collections.Immutable)**: Immutable collections such as `ImmutableList<T>`, `ImmutableDictionary<TKey, TValue>`, and `ImmutableHashSet<T>` are thread-safe by design and work well in scenarios where a collection is not expected to change, or you want to avoid side effects from mutations. They also eliminate the need for synchronization primitives (such as locks) when accessing these collections from multiple threads, reducing overhead and avoiding potential lock contention issues. When a new collection is created due to an operation (such as adding or removing elements), it reuses most of the existing structure instead of copying the entire collection. This can significantly reduce memory usage and the overhead of creating new collections, especially for large datasets.

The following is a benchmark comparison between these collection types, where we add elements and then iterate over them:

```
public class CollectionBenchmark
{
    private const int N = 1000;

    [Benchmark]
    public void AddToList()
    {
        List<int> list = new List<int>();
        for (int i = 0; i < N; i++)
        {
            list.Add(i);
        }
    }

    [Benchmark]
    public void AddToArray()
    {
        int[] array = new int[N];
        for (int i = 0; i < N; i++)
        {
            array[i] = i;
        }
    }

    [Benchmark]
    public void AddToConcurrentBag()
    {
        ConcurrentBag<int> bag = new ConcurrentBag<int>();
        for (int i = 0; i < N; i++)
        {
            bag.Add(i);
        }
    }

    [Benchmark]
    public void AddToImmutableList()
    {
        ImmutableList<int> immutableList = ImmutableList<int>.Empty;
```

```
        for (int i = 0; i < N; i++)
        {
            immutableList = immutableList.Add(i);
        }
    }

    [Benchmark]
    public void IterateList()
    {
        List<int> list = new List<int>(Enumerable.Range(0, N));
        foreach (var item in list) { }
    }

    [Benchmark]
    public void IterateArray()
    {
        int[] array = Enumerable.Range(0, N).ToArray();
        foreach (var item in array) { }
    }

    [Benchmark]
    public void IterateConcurrentBag()
    {
        ConcurrentBag<int> bag = new ConcurrentBag<int>
        (Enumerable.Range(0, N));
        foreach (var item in bag) { }
    }

    [Benchmark]
    public void IterateImmutableList()
    {
        ImmutableList<int> immutableList =
        ImmutableList.CreateRange(Enumerable.Range(0, N));
        foreach (var item in immutableList) { }
    }
}
```

Let us execute the code using the following command:

```
dotnet run -c release
```

Figure 3.2 shows the result and shows that the performance in a single-threaded application favors the standard List<T> and Array data types:

```
// * Summary *

BenchmarkDotNet v0.13.12, Windows 11 (10.0.22631.3296/23H2/2023Update/SunValley3)
Intel Core i7-9750H CPU 2.60GHz, 1 CPU, 12 logical and 6 physical cores
.NET SDK 8.0.200
  [Host]     : .NET 8.0.2 (8.0.224.6711), X64 RyuJIT AVX2
  DefaultJob : .NET 8.0.2 (8.0.224.6711), X64 RyuJIT AVX2


| Method             | Mean         | Error       | StdDev      |
|------------------- |-------------:|------------:|------------:|
| AddToList          |   2,149.3 ns |    42.33 ns |    37.52 ns |
| AddToArray         |     471.6 ns |    14.08 ns |    39.94 ns |
| AddToConcurrentBag |  18,504.0 ns |   235.36 ns |   208.64 ns |
| AddToImmutableList | 165,444.0 ns | 3,254.62 ns | 3,874.39 ns |
| IterateList        |   1,117.5 ns |    21.97 ns |    20.55 ns |
| IterateArray       |     537.6 ns |    11.52 ns |    33.60 ns |
| IterateConcurrentBag | 15,657.1 ns |  303.88 ns |   284.25 ns |
| IterateImmutableList | 22,489.2 ns |  381.07 ns |   318.21 ns |

// * Hints *
Outliers
  CollectionBenchmark.AddToList: Default           -> 1 outlier  was  removed (2.47 us)
  CollectionBenchmark.AddToArray: Default          -> 7 outliers were removed (615.27 ns..725.49 ns)
  CollectionBenchmark.AddToConcurrentBag: Default  -> 1 outlier  was  removed (19.22 us)
  CollectionBenchmark.IterateList: Default         -> 2 outliers were removed (1.23 us, 1.23 us)
  CollectionBenchmark.IterateArray: Default        -> 2 outliers were removed (660.72 ns, 661.54 ns)
  CollectionBenchmark.IterateConcurrentBag: Default -> 1 outlier  was  removed, 2 outliers were detected (14.92 us, 16.27 us)
  CollectionBenchmark.IterateImmutableList: Default -> 2 outliers were removed (23.67 us, 24.31 us)

// * Legends *
  Mean   : Arithmetic mean of all measurements
  Error  : Half of 99.9% confidence interval
  StdDev : Standard deviation of all measurements
  1 ns   : 1 Nanosecond (0.000000001 sec)

// ***** BenchmarkRunner: End *****
Run time: 00:03:40 (220.59 sec), executed benchmarks: 8

Global total time: 00:04:01 (241.36 sec), executed benchmarks: 8
// * Artifacts cleanup *
Artifacts cleanup is finished
```

Figure 3.2 – Result of a benchmark test for different collection types

Immutable and concurrent collections often show slower performance and higher memory usage for add operations due to the need to create a new collection each time. In contrast, List<T> and arrays might perform better due to direct element insertion since arrays typically offer the best iteration performance due to their contiguous memory layout, which is cache-friendly, and List<T> follows closely.

Generally, we prefer arrays or List<T> for lower memory overhead in single-threaded applications or when collection mutation is localized to a single thread. Use concurrent collections in multithreaded scenarios where operations on the collection are frequent and concurrent, accepting the trade-off of higher memory usage for out-of-the-box thread safety. We will explore these multithreaded scenarios later.

Now, let us review the concept of pre-sizing our data structures.

Pre-sizing data structures

The concept of pre-sizing data structures in .NET refers to specifying a collection's initial size or capacity when it's created based on the expected number of elements it will hold. This practice can significantly improve performance, especially when you add many items to collections, such as lists, dictionaries, or hash sets.

When you add elements to a dynamically sized collection without specifying an initial capacity, the collection might need to increase its size multiple times to accommodate new elements. Each time the length is increased, the collection typically needs to allocate a new array larger than the previous one and then copy the elements from the old array to the new one. This process is called *resizing*, and it can be computationally expensive for the following reasons:

- Memory allocation for the new array is required each time
- Each element from the old array must be copied to the new array
- The old array becomes garbage that the GC must eventually reclaim

By pre-sizing, we allocate enough space for the expected number of elements beforehand, minimizing or eliminating the need for resizing. The following is an example of pre-sizing List<T>, even though this would have been a dynamic collection type:

```
int expectedItems = 1000;
List<int> numbers = new List<int>(expectedItems); // Pre-sized list
for (int i = 0; i < expectedItems; i++)
{
    numbers.Add(i);
}
```

The following code snippet shows a benchmark test in which we compare the allocation speed of a dynamically sized list with that of a fixed-sized list. It runs a benchmark comparison between adding elements to a dynamic list versus adding them to a pre-sized one:

```
public class ListBenchmark
{
    private const int NumberOfElements = 10000;

    [Benchmark]
    public void AddToDynamicList()
    {
        List<int> dynamicList = new List<int>(); // Dynamic list
        for (int i = 0; i < NumberOfElements; i++)
        {
            dynamicList.Add(i);
        }
    }
```

```
    }

    [Benchmark]
    public void AddToPreSizedList()
    {
        List<int> preSizedList = new List<int>(NumberOfElements);
        // Pre-sized list
        for (int i = 0; i < NumberOfElements; i++)
        {
            preSizedList.Add(i);
        }
    }
}
```

To execute, we reuse the following command.

```
dotnet run -c release
```

The output looks like this:

```
// * Summary *

BenchmarkDotNet v0.13.12, Windows 11 (10.0.22631.3296/23H2/2023Update/SunValley3)
Intel Core i7-9750H CPU 2.60GHz, 1 CPU, 12 logical and 6 physical cores
.NET SDK 8.0.200
  [Host]     : .NET 8.0.2 (8.0.224.6711), X64 RyuJIT AVX2
  DefaultJob : .NET 8.0.2 (8.0.224.6711), X64 RyuJIT AVX2


| Method            | Mean     | Error    | StdDev   |
|------------------ |---------:|---------:|---------:|
| AddToDynamicList  | 21.05 us | 0.274 us | 0.243 us |
| AddToPreSizedList | 14.81 us | 0.280 us | 0.525 us |

// * Hints *
Outliers
  ListBenchmark.AddToDynamicList: Default  -> 1 outlier  was  removed (25.06 us)
  ListBenchmark.AddToPreSizedList: Default -> 3 outliers were removed (16.36 us..17.07 us)

// * Legends *
  Mean   : Arithmetic mean of all measurements
  Error  : Half of 99.9% confidence interval
  StdDev : Standard deviation of all measurements
  1 us   : 1 Microsecond (0.000001 sec)

// ***** BenchmarkRunner: End *****
Run time: 00:01:14 (74.48 sec), executed benchmarks: 2

Global total time: 00:01:33 (93.24 sec), executed benchmarks: 2
// * Artifacts cleanup *
Artifacts cleanup is finished
```

Figure 3.3 – Benchmark results for adding items to a fixed and dynamic-sized list.

The next tip sees us using contiguous memory as often as we can.

Accessing contiguous memory

Contiguous memory allocation ensures that elements of a data structure are stored next to each other in memory. This can significantly improve cache locality and reduce the time it takes to read and write data. As we have already explored, arrays are the simplest form of contiguous memory allocation. They store elements of the same type in a continuous block of memory, allowing for fast access by index:

```
int[] numbers = new int[5] { 1, 2, 3, 4, 5 };

// Accessing an element
int thirdNumber = numbers[2]; // Access is O(1)
```

Span<T> and Memory<T> were introduced in .NET Core 2.1 and C# 7 to provide a type-safe way to represent contiguous memory regions. When used, they do not own the memory but rather provide a window into memory. This can be stack-allocated memory, managed arrays, or native memory. Span<T> is a stack-only type suitable for short-lived scenarios, while Memory<T> can be stored on the heap, making it ideal for async methods. The following is an example of using the Slice() method to view some elements from an array as a Span<T>:

```
int[] array = new int[] { 0, 1, 2, 3, 4, 5 };
Span<int> span = new Span<int>(array);

// Creating a slice of the array from index 1 to 3
Span<int> slice = span.Slice(1, 3);

foreach (var item in slice)
{
    Console.WriteLine(item); // Outputs 1, 2, 3
}
```

ArraySegment<T> is another method to view a segment of an array without allocating additional memory for the segment. Like Span<T> and Memory<T>, it allows for accessing contiguous memory efficiently, but it is a reference type and can be used where Span<T> cannot:

```
int[] numbers = { 1, 2, 3, 4, 5 };
ArraySegment<int> segment = new ArraySegment<int>(numbers, 1, 3);

foreach (int i in segment)
{
    Console.WriteLine(i); // Outputs 2, 3, 4
}
```

By utilizing data structures and types that use contiguous memory, developers can achieve significant performance improvements through improved cache locality and reduced memory access times.

Now, we have explored scenarios that help us to write memory-efficient applications. We have also established that not everything that shines is gold, and each tool has a scenario where it works better and one where it is not the best choice. We must diligently understand the algorithm we are implementing and the factors that might influence its performance. With this insight, we can be more poignant in deciding which tools to use.

Next, we will explore hands-on methods for handling large objects and arrays.

Handling large objects and arrays

This topic requires careful consideration due to the impact that collections of large objects can have on application performance and memory usage. Large objects are typically defined as 85,000 bytes or larger. To improve the efficiency of garbage collection, these objects are allocated in a particular area of the heap known as the **Large Object Heap (LOH)**.

Objects on the LOH are collected during a Gen 2 collection. Because Gen 2 collections are less frequent, large objects can remain in memory longer, increasing memory usage. Since the LOH is not compacted, the heap can become fragmented over time, potentially leading to out-of-memory exceptions or increased memory usage because of the inability to utilize free spaces efficiently.

The garbage collector doesn't compact the LOH because moving large objects around it is costly. Instead, it removes the memory occupied by large objects when they are no longer in use. As a result, over time, memory holes are created in the LOH, or the memory becomes fragmented. Memory fragmentation adversely affects your application's performance and scalability.

Although the garbage collector doesn't compact the LOH, you can still compact the LOH explicitly using the following piece of code:

```
GCSettings.LargeObjectHeapCompactionMode =
GCLargeObjectHeapCompactionMode.CompactOnce;
GC.Collect();
```

Although you can compact the LOH programmatically, it's recommended that you avoid using the LOH as much as possible. While avoiding the LOH entirely may not be feasible for all applications, the following strategies can help reduce reliance on it, mitigate memory fragmentation, and improve overall application performance.

Each application is unique, so it's important to profile and test changes to understand their impact on performance and memory usage. Let us discuss using smaller objects where possible.

Using smaller objects

Split large data structures into smaller chunks if feasible. For example, consider using a list of smaller arrays instead of a single large array. This can help keep the individual arrays below the LOH threshold. This approach is often used in scenarios where you might be tempted to use a very large array, such as in buffering large amounts of data for processing.

Suppose you're implementing a feature that processes a large dataset. Instead of a single large array, you could use a list of arrays, each of a manageable size:

```csharp
using System;
using System.Collections.Generic;

public class SmallObjectManager
{
    // Each array size is set to a value that won't be allocated
    // on the LOH.
    private const int MaxArraySize = 20_000;
    // size * sizeof(int) < 85,000 bytes for int
    private List<int[]> arrays = new List<int[]>();

    public void AddData(IEnumerable<int> data)
    {
        int[] currentArray = null;
        int currentIndex = 0;

        foreach (var item in data)
        {
            // Allocate a new array when needed.
            if (currentArray == null || currentIndex >= MaxArraySize)
            {
                currentArray = new int[MaxArraySize];
                arrays.Add(currentArray);
                currentIndex = 0;
            }

            currentArray[currentIndex++] = item;
        }
    }

    public IEnumerable<int> GetData()
    {
        foreach (var array in arrays)
        {
```

```csharp
            foreach (var item in array)
            {
                yield return item;
            }
        }
    }
}

public class Program
{
    public static void Main()
    {
        SmallObjectManager manager = new SmallObjectManager();

        // Example: Adding a large amount of data in chunks
        manager.AddData(GenerateLargeDataSet());

        // Retrieve and process the data
        foreach (var item in manager.GetData())
        {
            // Process each item
            Console.WriteLine(item);
        }
    }

    // Simulate generating a large set of data
    private static IEnumerable<int> GenerateLargeDataSet()
    {
        const int largeSize = 100_000;
        for (int i = 0; i < largeSize; i++)
        {
            yield return i;
        }
    }
}
```

The SmallObjectManager class internally manages a list of int arrays (List<int []>), ensuring that each array stays below the LOH threshold. When data is added, it checks whether a new array needs to be allocated. The GetData method iterates over this segmented storage as a single collection, abstracting away the underlying complexity. By keeping individual arrays below the threshold for the LOH, this strategy reduces the risk of memory fragmentation and the associated performance issues.

We have already discussed why strings can be dangerous memory constructs. Review their potential for LOH allocation and how we can implement efficient string manipulation.

Optimizing string usage

Less seasoned developers don't often realize that a string value in C# is an immutable collection of characters. This means that each time we seem to modify a string, we force the creation of a new string (or character array) in memory and leave the previous version for garbage collection. If a string value is large enough, it may end up in the LOH, and we can appreciate, by now, how that might affect your application's performance over time.

Just as we apply special considerations to general collection types, we must be careful when we use strings in our applications.

The first tip is to avoid concatenations. Each concatenation creates a new string object, which can quickly lead to large object allocations if the resulting string exceeds the LOH threshold.

The following code snippet shows a standard concatenation but a very harmful operation:

```
string result = "";
for (int i = 0; i < 10000; i++)
{
    result += "some text ";
}
```

Concatenating strings, especially in a loop, can lead to excessive memory usage and poor performance. The `StringBuilder` class is designed to handle string concatenation scenarios efficiently:

```
StringBuilder builder = new StringBuilder();
for (int i = 0; i < 10000; i++)
{
    builder.Append("some text ");
}
string result = builder.ToString();
```

String interning is another method that can reduce the memory used for string storage and can save memory when many identical strings are used by storing only one copy of each distinct string value. Interning is useful when you know you'll deal with many duplicate strings. Still, it should be used judiciously because the garbage collector does not collect the interned strings until the application domain is unloaded:

```
string a = string.Intern("Hello, world!");
string b = string.Intern("Hello, world!");

bool areSameReference = object.ReferenceEquals(a, b); // True
```

For operations involving large texts (e.g., file processing or network communication), consider using **character arrays** or the `Span<char>` type instead of string objects for mutable operations. This avoids creating intermediate string objects that could be allocated on the LOH:

```
char[] largeTextValue = new char[8192]; // 16KB buffer for characters
// Fill or modify the buffer

// use the Span<char>
char[] source = new char[8192]; // Large char array
Span<char> span = new Span<char>(source);
// Use span.Slice, span.CopyTo, etc., to work with portions of the
// buffer
```

Optimizing string usage in .NET involves being mindful of how and when strings are created, manipulated, and stored. Techniques such as string builder, string interning, and character arrays can reduce the impact on memory usage and performance, especially concerning the LOH. Remember, each optimization technique should be applied based on your application's specific needs and context, and the benefits should be measured through profiling and testing. Now, let us summarize this chapter.

Summary

This chapter explored several dimensions of memory allocation and optimization concepts. The recurring theme has been that different constructs bring different benefits to our algorithms and applications. We began by discussing the basics of memory allocation in Java, C, and C++ since these are foundational languages and worthy rivals to C#. We saw where C and C++ place the responsibility in the developer's hands to handle memory allocation and deallocation. In contrast, Java and .NET employ an automatic clean-up process and are guided by similar philosophies.

We also delved into various .NET data structures, such as arrays, lists, dictionaries, and more specialized collections such as `Span<T>`, `Memory<T>`, and concurrent collections. We explored how choosing the proper data structure for the task can significantly impact memory usage and application performance.

Memory allocation and data structures are fundamental to .NET programming, influencing applications' performance and reliability. By applying the principles and practices outlined in this chapter, developers can write more efficient, scalable, and maintainable .NET code. Understanding the nuances of .NET's memory management model allows developers to optimize their applications for the best possible performance and memory footprint, ensuring a smooth and responsive user experience. In the next chapter, we will take a closer look at how we can manage resources and avoid memory leaks.

4

Memory Leaks and Resource Management

It is well documented that memory management is critical in ensuring that applications run efficiently, smoothly, and without undue resource consumption. The need to ensure that this is adequately handled has led to automated management systems, as seen in development frameworks such as Java and .NET.

Understanding and managing memory and resources is vital for building robust, scalable, and performance-optimized applications. Memory leaks occur when a program fails to release memory that is no longer needed, gradually consuming more system resources and potentially leading to application slowdowns or crashes.

This chapter delves into the nuances of memory leaks and resource management in .NET development. It comprehensively explores their significance, the advantages and disadvantages of various management techniques, and the best practices for developers. We will explore key concepts and mitigation methods such as the following:

- Identifying memory leaks

- Memory analysis in .NET

- Best practices for avoiding memory leaks

Memory leaks remain a valid concern in .NET development despite the framework's sophisticated **garbage collection (GC)** mechanism. Developers must deeply understand how .NET manages memory and resources to prevent leaks, ensure efficient resource utilization, and maintain application performance.

By adhering to the guidelines that will be outlined in this chapter and maintaining a vigilant approach to memory and resource management, .NET developers can build applications that are performant, scalable, robust, and efficient in their resource utilization. This chapter aims to arm developers with the knowledge and tools necessary to navigate the challenges of memory leaks and resource management in .NET development, fostering a deeper understanding of the framework's capabilities and limitations.

Technical requirements

- Visual Studio 2022 Studio, which you can find at `https://visualstudio.microsoft.com/vs/community/`.

- Visual Studio Code, which can be found at `https://code.visualstudio.com/`.

- .NET 8 SDK, which you'll find at `https://dotnet.microsoft.com/en-us/download/visual-studio-sdks`.

Identifying memory leaks

Before exploring detection methods, it's essential to understand memory leaks and why they occur. A memory leak occurs when a program allocates memory for use but fails to release it after it's no longer needed. Over time, these leaks can accumulate, resulting in increased memory usage and application or system crashes due to resource exhaustion.

As applications grow in complexity and size, efficient memory management becomes a cornerstone of software development. Memory leaks, which are often subtle and gradual, can significantly degrade the performance and reliability of applications if left unchecked. Identifying these leaks early and effectively is crucial in maintaining optimal operation and ensuring a seamless user experience. This chapter aims to demystify the process of detecting memory leaks within application development, offering insights into tools, techniques, and methodologies designed to unearth and address these elusive issues.

To investigate potential memory leaks, it's necessary to examine the application's memory heap to scrutinize its contents. By observing how objects relate to each other, you can formulate hypotheses regarding why memory isn't being released as expected. Memory leaks are particularly pervasive in languages and environments where the developer manages manual memory. However, they can also occur in managed environments because references are kept alive unnecessarily, preventing garbage collectors from reclaiming the memory.

Common causes of memory leaks

Memory leaks can subtly erode the performance of applications, leading to issues that are often difficult to diagnose and resolve. Understanding the common causes of memory leaks is crucial for developers to prevent them and ensure efficient, reliable software. Here are some of the most prevalent causes of memory leaks across various programming environments:

- **Uncollected garbage**: In managed languages such as Java or C#, where GC is used, memory leaks often occur not because memory is not freed but because objects are unintentionally retained. This retention prevents the garbage collector from reclaiming the memory. Common culprits include static fields, observers or event listeners that are not properly removed, and data structures that grow indefinitely.

- **Improper resource management**: In managed and unmanaged environments, failing to release resources such as file handles, database connections, or network sockets can lead to memory leaks.

- **Closures**: In languages that support closures (functions that can capture and retain state from their surrounding scope), keeping more objects in memory is easier than intended. If a closure captures a large object or a complex scope but only uses a small part of it, the entire captured scope remains in memory, which can lead to a memory leak if it prevents the GC from cleaning up the reference.

- **Circular references**: Circular references can lead to memory leaks, particularly in languages with automatic GC. This situation occurs when two or more objects reference each other, creating a cycle that can prevent the garbage collector from determining that they are no longer in use and should thus be collected.

- **Global variables**: Overuse of global variables or extensive caching mechanisms without proper eviction policies can cause memory to be consumed progressively over time. Global variables remain in memory for the lifetime of an application. If they reference large data structures or objects, they can significantly contribute to memory leaks.

- **Improper use of data structures**: Certain data structures, especially those that dynamically grow, can cause memory leaks if not used carefully. For example, a dynamically resizing array that keeps increasing in response to application events without being trimmed or cleared can lead to excessive memory use.

- **Third-party code and libraries**: Dependencies on third-party libraries or frameworks can also introduce memory leaks. If these libraries do not correctly manage memory or resources, or if they retain more objects in memory than necessary, they can cause leaks in applications that use them. Regular updates and careful selection of well-maintained libraries can mitigate this risk.

- **Improper in-memory caching**: In-memory caching can significantly improve the performance of an application by reducing the need for repetitive data fetching and processing. However, if not managed properly, it can also lead to memory leaks under the following conditions:

  - A size limit and expiration policy are not configured. The cache may grow without bounds and eventually consume all available memory.

  - Cached items are no longer needed but are not explicitly removed.

  - A cache retains a strong reference to an entry, preventing GC even if it is no longer used elsewhere in the application.

Now that we have explored some of the more common causes of memory leaks, let's examine mitigation methods.

Mitigating memory leaks in .NET applications requires a comprehensive approach that involves understanding the sources of leaks, implementing best practices, and using appropriate tools. Regularly reviewing and refactoring your code, along with educating your development teams, will help maintain high standards of memory management throughout your application's life cycle. Let us take a closer look at how code reviews can assist in mitigating leaks.

Code review

Code reviews bring developers with varying levels of expertise and perspectives together. This collective wisdom is particularly beneficial in identifying memory leaks, as more experienced developers can share insights into common pitfalls and indicators of potential leaks that less experienced team members might overlook.

Regular code reviews, both manual and automated, can help identify potential sources of memory leaks. Reviewers can look for common leak patterns, such as missing deallocators or unmanaged resource allocation without corresponding release logic. They are a critical checkpoint in the software development life cycle, offering a structured opportunity to catch a wide range of issues including the subtle and often elusive memory leaks. When conducted effectively, code reviews can illuminate problematic patterns and practices that may lead to memory leaks, even before the code is executed.

In the context of .NET development, code reviews are crucial in identifying and preventing memory leaks. These can occur even in a managed memory environment such as .NET due to improper resource use or patterns that prevent GC. Even though .NET's garbage collector is designed to free memory allocated to objects that are no longer in use, certain coding patterns can inadvertently keep objects alive, thus preventing the garbage collector from reclaiming their memory. Some coding patterns that are well-known for their potential to create memory leaks include, but are not limited to, the following:

- **Unclosed resources (such as files or network connections)**: It is essential to ensure that the resources consumed for these connections are freed as soon as the operation is completed.

- **Undetached event listeners**: In .NET, it's easy to create memory leaks through event handlers by subscribing to events and failing to unsubscribe. Reviewers should ensure that event handlers are properly detached, especially for long-lived publishers.

- **Misuse of static variables**: Static fields or properties referencing large objects or collections can keep those objects alive indefinitely. Reviewers can look for unnecessary static references that could be converted to instance-level or more transient scopes.

These are some of the points that a code reviewer should be keen on identifying. At a high level, this may seem like trying to find a needle in a haystack, but more experienced developers will be able to identify code smells that can lead to these issues. Once these issues are identified, a good mentor will address them and share insights on potential problems that the now-corrected code might have had, as well as the benefits of the refactored code, with fellow developers.

The following is an example of a controller in a .NET Core API. It contains a static list, which is a potential memory leak waiting to happen:

```
namespace API.Chapter04.Controllers;

[ApiController]
[Route("[controller]")]
public class WeatherForecastController(IWeatherService weatherService)
 : ControllerBase
{
    private static List<string> _leakyList = new List<string>();

    [HttpGet]
    public IEnumerable<string> Get()
    {
        // Simulate memory leak
        _leakyList.Add(new string('a', 1024 * 1024));
        // Add 1MB to the list

        return new string[] { "Data added to the leaky list.
        Total items: " + _leakyList.Count };
    }
}
```

This code introduces a memory leak by adding a large string (1 MB) to a static list each time the Get method is called. The list keeps growing with no mechanism to clear or remove the items, simulating a memory leak.

While conducting a code review, always seek to identify static fields and properties that might hold references to large objects or object graphs. If a static reference is found, question its necessity and suggest alternatives such as **dependency injection (DI)** or instance-based management where appropriate.

After the code review identifies the memory leak in the given .NET Core API, refactoring involves modifying the application to properly manage memory and ensuring that resources are not unnecessarily retained. Some measures to address this could include the following:

- **Use a local variable**: Remove the static list and use local variables or instance-level storage as needed, ensuring that they are collected by GC when they are no longer in use.

- **Limit the list size**: If the static variable serves a functional purpose, decide on a reasonable list size limit. Once that limit is reached, you can start removing old items before adding new ones or clearing the list periodically.

- **Use a more optimized data structure**: If the static data must be preserved, consider using more memory-efficient data structures or techniques that better suit your requirements, such as a **fixed-size queue**.

After refactoring this method to use a scoped list variable instead of a static one, the object is destroyed after each request cycle. The resulting endpoint looks like this:

```
namespace API.Chapter04.Controllers;

[ApiController]
[Route("[controller]")]
public class WeatherForecastController(IWeatherService weatherService)
: ControllerBase
{
    private List<string> _properList = new List<string>();

    [HttpGet]
    [Route("GetRefactored")]
    public IEnumerable<string> GetRefactored()
    {
        // Simulate memory leak
        _properList.Add(new string('a', 1024 * 1024));
        // Add 1MB to the list

        return new string[] { "Data added to the proper list.
        Total items: " + _properList.Count };
    }
}
```

After these changes, another review should be conducted to ensure that no other code smells are introduced.

Code reviews are essential in software development, providing a structured process for peers to evaluate each other's code. They play a crucial role in maintaining code quality, ensuring adherence to standards, and improving team knowledge. Some tips to ensure that a proper code review is performed include the following:

- Use a checklist with memory management items to ensure consistent review practices
- Collaborate so reviewers and authors can discuss potential memory leaks and remediation strategies
- Use code reviews to share knowledge about common memory leak patterns and best practices for memory management
- Integrate static analysis tools that can automatically detect some common patterns that lead to memory leaks and highlight them during the code review process

Stress testing is another suitable method for finding memory leaks. Let us review this next.

Stress testing

Stress testing is a powerful development technique. It can uncover memory leaks that might not be evident during standard testing or at lower application usage levels. This approach is crucial for identifying memory leaks, as these often accumulate over time or under conditions of intense resource demand. Subjecting applications to stress testing, whereby the system is pushed to its operational limits, can help reveal memory leaks that might not be evident under normal operating conditions. Monitoring memory usage under stress can highlight unusual patterns or unexpected growth in memory consumption.

Memory leaks in .NET applications might not always be immediately apparent, especially if they are slow and accumulate over time. A gradual increase in memory usage might go unnoticed during regular testing or initial deployment phases. Real-world applications often face variable loads, with peak times significantly stressing system resources. Stress testing mimics these scenarios, providing insights into how the application manages memory under heavy loads.

While the garbage collector is designed to manage memory automatically, inefficient memory usage patterns or misuse of resources can hinder its effectiveness. Stress testing can reveal scenarios wherein the garbage collector cannot reclaim memory efficiently, possibly due to large object heap fragmentation, excessive pinning of memory, or improper disposal patterns. Observing the behavior of the garbage collector under stress can highlight areas where memory management can be optimized.

Some best practices that should govern your stress testing activities include the following:

- **Use realistic load scenarios**: Ensure that the stress tests mimic realistic user behaviors and load patterns as closely as possible

- **Monitor and measure**: Utilize .NET performance counters, memory profilers, and logging to monitor memory usage and system behavior under stress

- **Analyze GC metrics**: Pay attention to GC performance metrics to understand how effectively memory is being managed and reclaimed

- **Iterate and refine**: Stress testing should be an iterative process, whereby findings from each test are used to refine the application and improve its resilience and memory management.

Several capable tools exist to help us simulate application loads. The manual effort required to get them up and running is minimal, but the power of automation makes this effort worthwhile in the long run. We will use **Postman**, a powerful **HTTP, HTTPS, and SOAP/REST** client and testing tool, to perform stress testing on web applications.

You can acquire Postman here `https://www.postman.com/downloads/` and install it with a few simple steps. Once you have done that, you can configure a stress test with the following steps:

1. Create a new workspace, as seen in *Figure 4.1*.

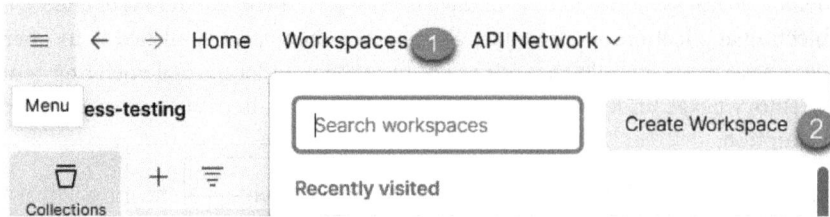

Figure 4.1 – Creating a new Postman workspace

2. Create a new collection, as seen in *Figure 4.2*, and stress-testing.

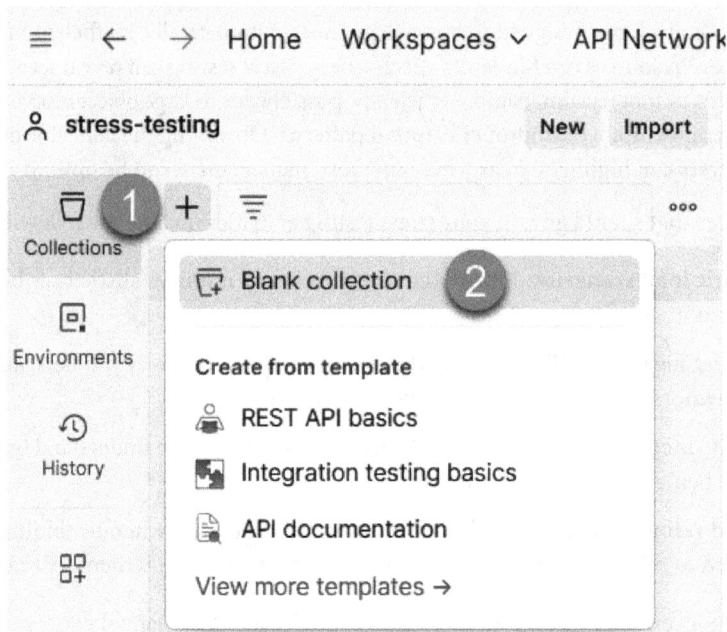

Figure 4.2 – Creating a new collection in Postman

Add a new request to the collection. This request will be the URL to the API endpoint you wish to test and save the change. In our case, we are using our `WeatherForecastController` with a memory leak, so the request URL will follow the format of `https://{baseURL}/weatherforecast`.

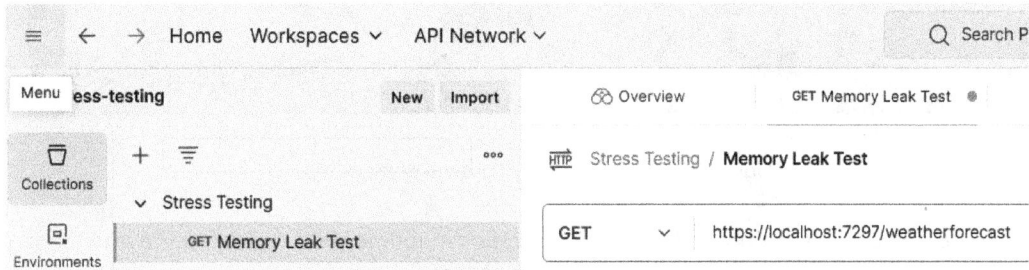

Figure 4.3 – Adding a new API request to the Postman collection

Right-click the newly created collection and choose **Run collection**. On the resulting screen, select the **Performance** tab and set the test parameters to your desired values. For this demo, we can reduce the test duration to **1 minute**. Once these values are confirmed, proceed to click **Run**. The steps are outlined in *Figure 4.4*.

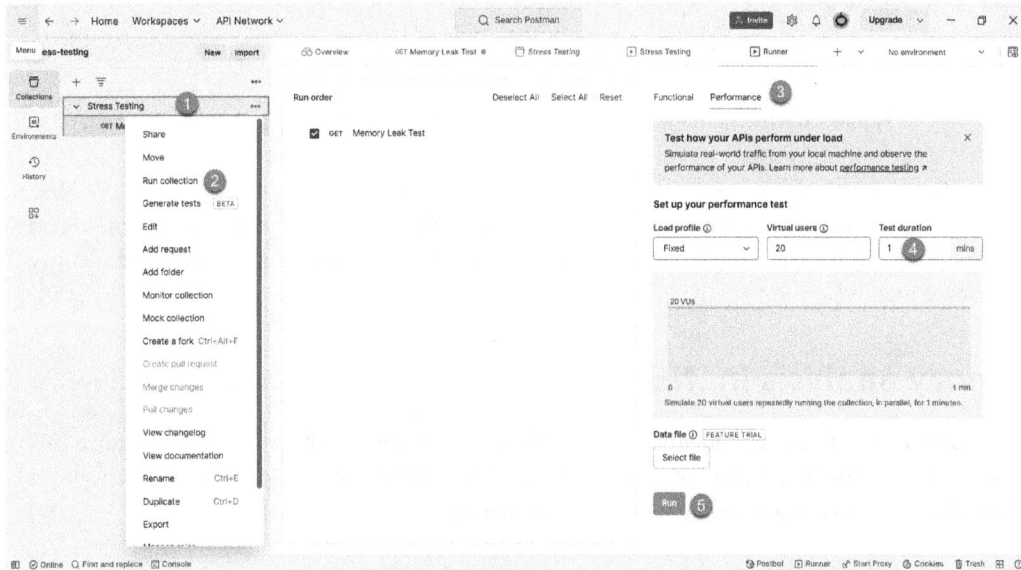

Figure 4.4 – The steps to configure a stress test in Postman

While this test runs, you can observe your memory usage, which will grow exponentially throughout the test. *Figure 4.5* shows the growth of memory usage. This view is from the **Visual Studio Diagnostic Tools**, which we will look at later in this chapter.

Figure 4.5 – The exponential growth of memory

We can see where GC events occur during the application's runtime but do not affect the amount of memory that is consumed. This evidence of a memory leak and mismanaged resources must be checked.

If we run this same stress test against our refactored endpoint, the memory usage results, as seen in *Figure 4.6*, will indicate that the garbage collector clears the memory before the subsequent request is made.

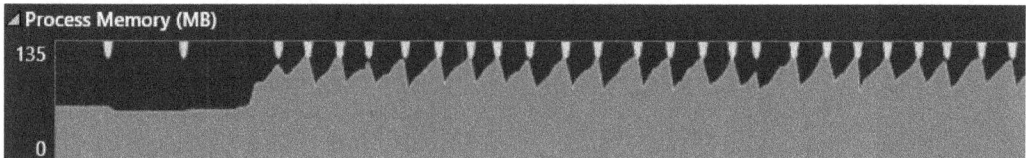

Figure 4.6 – The garbage collector recovering memory after each request

Stress testing is a critical component of the development life cycle for both web and desktop .NET applications. It ensures that applications are robust, scalable, and capable of handling real-world pressures. Now, let us review how to use built-in .NET tools to review memory usage and identify troublesome methods.

Memory analysis in .NET

Practical memory analysis helps identify leaks, understand memory consumption, and optimize GC. Various tools are available for memory analysis in .NET. We will seek to explore some of their features and usage, as well as the scenarios in which they are handy.

Suppose you are developing on a Windows OS. You can install the Windows **Assessment and Deployment Kit (ADK)**, which contains the Windows Assessment and Windows Performance toolkits. These allow you to assess the quality and performance of systems or components using both **Windows Performance Recorder (WPR)** and **Windows Performance Analyzer (WPA)**:

- **WPR**: This tool can collect system-wide performance data, including memory usage statistics. It also creates **Event Tracing for Windows (ETW)** recordings, which can run from a **user interface (UI)** or the command line.

- **WPA**: This tool analyzes performance recordings, helping developers understand memory patterns and identify leaks. It combines a robust UI with extensive graphing capabilities and data tables that support pivoting and full-text search capabilities. WPA provides an **Issues** window to explore the root cause of any identified.

Outside of acquiring these additional tools, we may rely on Visual Studio to provide real-time diagnostics during an application debugging session. This is especially useful since it helps us view potential memory leaks during development. We will explore this tooling next.

Visual Studio diagnostic tools

Visual Studio integrates debugger and diagnostic tools to provide a way to inspect memory usage during development. Developers can view real-time memory consumption and take snapshots of the memory heap to compare and analyze objects' memory allocation and GC.

There is the **Memory Usage** tool, which allows you to take memory snapshots at any point, but using the Visual Studio debugger provides enhanced control over your application's execution when probing performance issues. Employing debugger functionalities such as setting breakpoints, stepping through code, or using the **Break All** feature helps you concentrate on the most pertinent code paths. This targeted approach minimizes distractions from irrelevant code sections and drastically shortens the time needed to pinpoint and resolve issues.

Let us conduct an exercise wherein we run our leaky API endpoint in debugging mode and observe the visual cues that the diagnostic tools provide. We can begin by setting a breakpoint at the beginning of the function and another after the point where the suspected memory leak occurs (as seen in *Figure 4.7*). We will then execute a GET request using Postman.

```
12          [HttpGet]
            0 references
13     ∨    public IEnumerable<string> Get()
14          {
15              // Simulate memory leak
16              _leakyList.Add(new string('a', 1024 * 1024)); // Add 1MB to
                the list

17
18 ⑨         return new string[] { "Data added to the leaky list. Total
                items: " + _leakyList.Count };

19          }
```

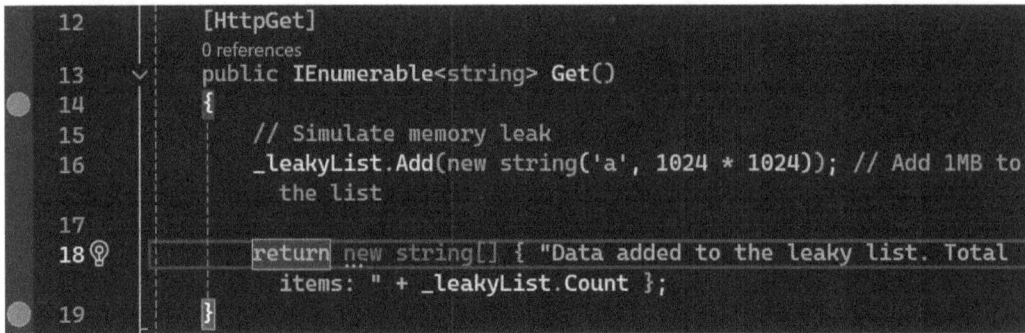

Figure 4.7 – The breakpoints on the leaky API endpoint

Run the program in debugging mode, and the **Diagnostic Tools** window should appear automatically unless you have hidden the pane or turned it off. To re-enable it, go to **Debug | Windows | Show Diagnostic Tools** in your context menu. The steps are also outlined in *Figure 4.8*.

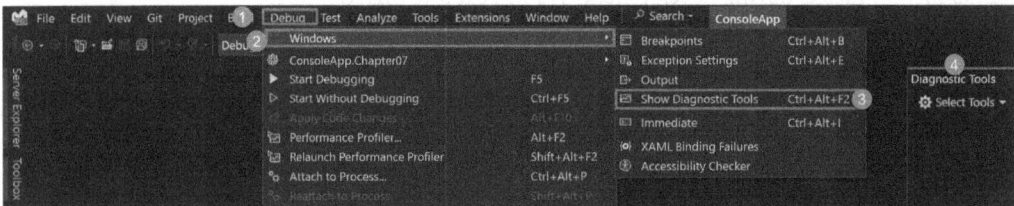

Figure 4.8 – Re-enabling the Diagnostic Tools window

This window will show us a few graphs to view **Events**, **Process Memory** (as seen in *Figures 4.5* and *4.6*), and **CPU usage**. These graphs can be handy for real-time performance tracking. Below the graphs are several tabs, including one labeled **Memory Usage**. From the **Memory Usage** tab, we can take a snapshot of the heap and objects when the application hits a breakpoint. The most important columns in the **Memory Usage** tab display are the following:

- **Objects (Diff)** (.NET): This displays the number of objects in .NET or native memory when the snapshot was taken

- **Heap Size (Diff)**: This displays the number of bytes in the .NET and native heaps

To begin this analysis, we will take a snapshot of the memory when the first breakpoint is hit and then again when the second breakpoint is hit. We can then compare the values to determine the memory usage growth during the operation. Your results should look like *Figure 4.9*.

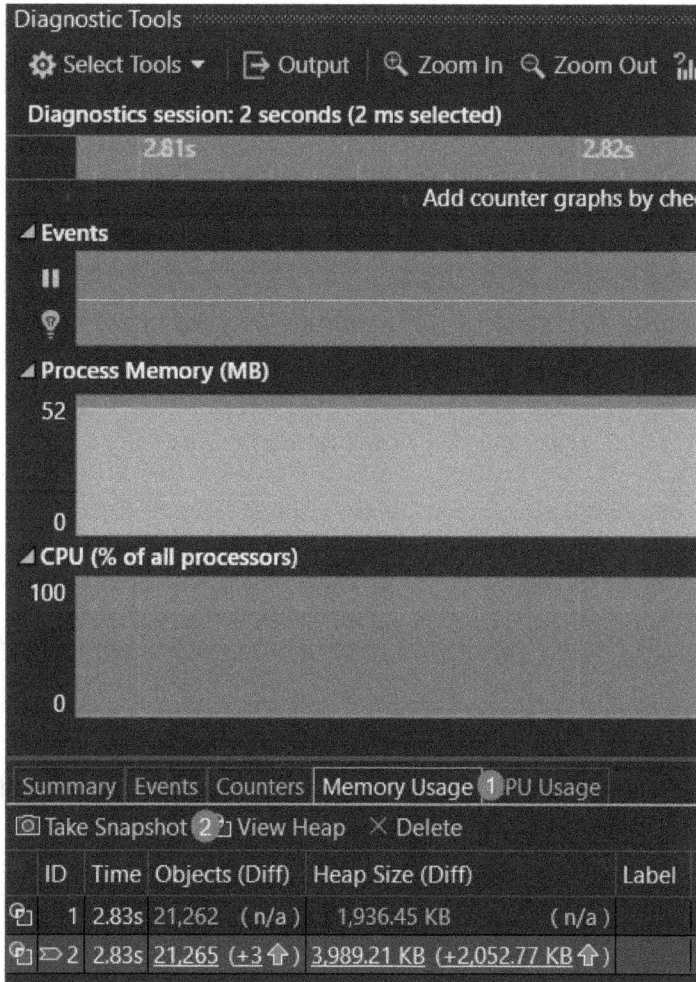

Figure 4.9 – The diagnostics pane with memory usage snapshots

We can observe that memory usage increased substantially between the breakpoints, but we must conduct several tests to conclude whether there is a memory leak in this operation. We can repeat these steps several more times. By taking additional snapshots, we can observe that the heap size is continuously growing. *Figure 4.10* shows that the heap size metric does not reduce after additional requests.

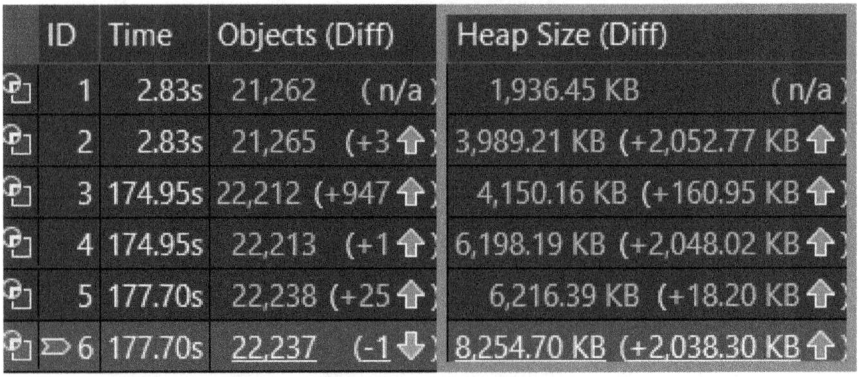

	ID	Time	Objects (Diff)	Heap Size (Diff)
	1	2.83s	21,262 (n/a)	1,936.45 KB (n/a)
	2	2.83s	21,265 (+3 ⬆)	3,989.21 KB (+2,052.77 KB ⬆)
	3	174.95s	22,212 (+947 ⬆)	4,150.16 KB (+160.95 KB ⬆)
	4	174.95s	22,213 (+1 ⬆)	6,198.19 KB (+2,048.02 KB ⬆)
	5	177.70s	22,238 (+25 ⬆)	6,216.39 KB (+18.20 KB ⬆)
	6	177.70s	22,237 (-1 ⬇)	8,254.70 KB (+2,038.30 KB ⬆)

Figure 4.10 – The increasing heap size value

To view the difference between the current and previous snapshots, click the change link to the left of the red arrow. A red arrow indicates an increase in memory usage and a green arrow indicates a decrease. As seen in *Figure 4.11*, the String object type uses most of the memory for a particular snapshot.

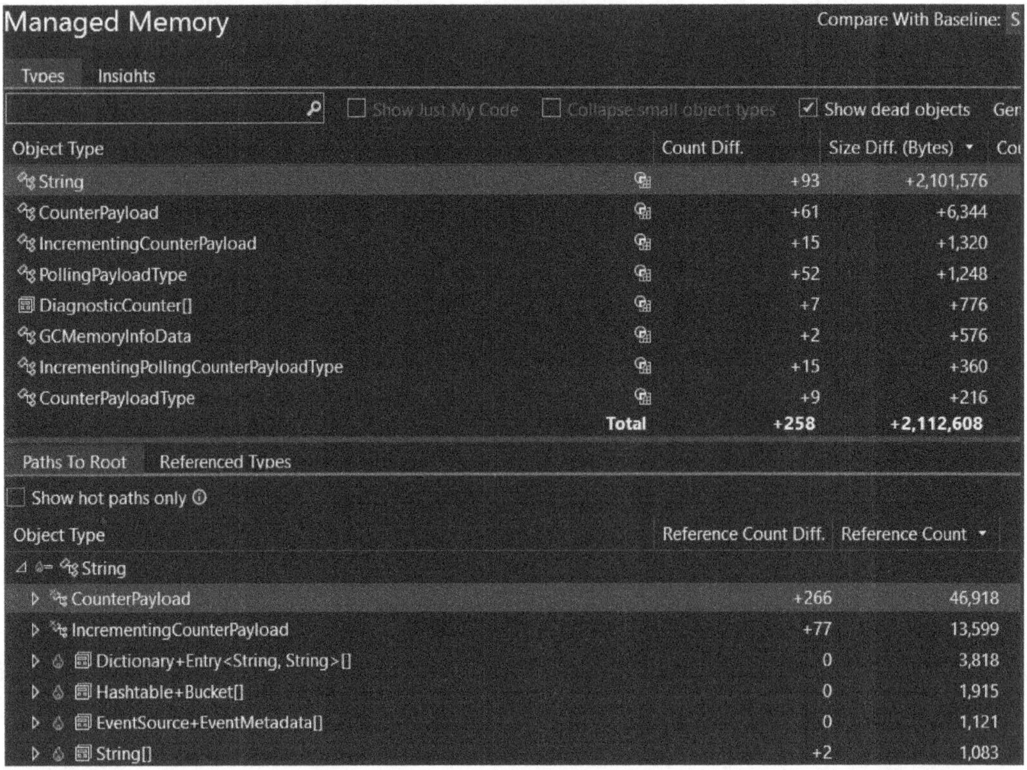

Object Type		Count Diff.	Size Diff. (Bytes) ▾	Co
String		+93	+2,101,576	
CounterPayload		+61	+6,344	
IncrementingCounterPayload		+15	+1,320	
PollingPayloadType		+52	+1,248	
DiagnosticCounter[]		+7	+776	
GCMemoryInfoData		+2	+576	
IncrementingPollingCounterPayloadType		+15	+360	
CounterPayloadType		+9	+216	
Total		**+258**	**+2,112,608**	

Object Type	Reference Count Diff.	Reference Count ▾
△ ⬠ String		
▷ CounterPayload	+266	46,918
▷ IncrementingCounterPayload	+77	13,599
▷ ◊ Dictionary+Entry<String, String>[]	0	3,818
▷ ◊ Hashtable+Bucket[]	0	1,915
▷ ◊ EventSource+EventMetadata[]	0	1,121
▷ ◊ String[]	+2	1,083

Figure 4.11 – The analysis of a snapshot

We can now double-click the String object type in the list to view instances where it was used. This will give us a more detailed view of how and when the String objects were instantiated. This instance view can be seen in *Figure 4.12*.

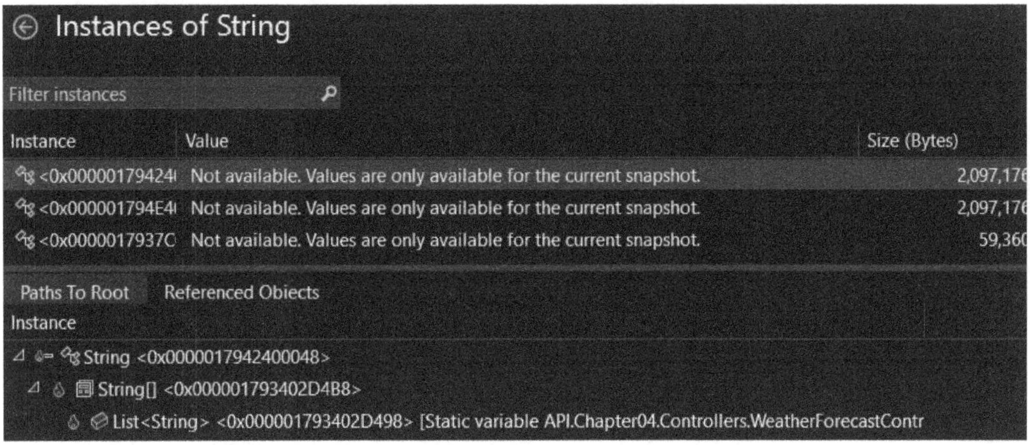

Figure 4.12 – The analysis of a snapshot

This robust diagnostic tooling allows us to write better applications and find leaks before shipping our software. Unfortunately, Visual Studio is currently only available for Windows. Given that .NET is cross-platform, we need to be able to perform analysis regardless of the development environment. Next, we will explore how to do this with the .NET CLI.

Analysis with the .NET CLI

The .NET SDK is cross-platform, which removes the limitations of some Windows-only development activities. We can use the .NET CLI to access the underlying functionality that all other tools are built on top of.

The .Net CLI provides the **dotnet-counters** tool to help us confirm a potential memory leak in our application. dotnet-counters serves as a tool for immediate health checks and preliminary performance analysis. It monitors values from performance counters, which are made available through the **EventCounter API** or the **Meter API**. This allows for swift observation of various metrics, such as CPU usage or the frequency of exceptions in your .NET Core application, helping to identify any potential issues immediately before engaging in deeper performance investigations with tools such as PerfView or dotnet-trace.

We need to install this tool or ensure that you have updated it to the latest version. To do this, open a terminal window and run the following command:

```
dotnet tool install --global dotnet-counters
```

Using the previous stress testing method, we will run our API and continuously call our API endpoint with the memory leak. While this stress test is in progress, check managed memory usage with the `dotnet-counters` tool using the following command:

```
dotnet-counters ps
```

You will get an output listing several running processes. In *Figure 4.13*, you will see a specific line item for our running app.

```
PS                dotnet-counters ps
  3916  Agent.Listener        C:\agent\new-aget\bin\Agent.Listener.exe        n\Agent.Listener.exe" run --startuptype service
  9300  Agent.Listener        C:\agent\bin\Agent.Listener.exe                 n\Agent.Listener.exe" run --startuptype service
 31876  API.Chapter04         PI.Chapter04\bin\Debug\net8.0\API.Chapter04.exe  .Chapter04\bin\Debug\net8.0\API.Chapter04.exe"
 32392  PhoneExperienceHost   77.0_x64__8wekyb3d8bbwe\PhoneExperienceHost.exe  ienceHost.exe" -ComServer:Background -Embedding
```

Figure 4.13 – The process running in memory

Now that you have the process ID, we can check managed memory usage with the `dotnet-counters` tool and the `--refresh-interval` parameter, which specifies the number of seconds between refreshes. This will give us live output during the runtime of the stress test:

```
dotnet-counters monitor --refresh-interval 1 -p 31876
```

This live output will look something like what is seen in *Figure 4.14*. f we focus on the line item **GC Heap Size (MB)**, we can observe the memory usage. By watching this, you can safely determine whether memory is growing or leaking.

```
Press p to pause, r to resume, q to quit.
    Status: Running

Name                                                        Current Value
[System.Runtime]
    % Time in GC since last GC (%)                                      0
    Allocation Rate (B / 1 sec)                               42,043,264
    CPU Usage (%)                                                       0
    Exception Count (Count / 1 sec)                                    0
    GC Committed Bytes (MB)                                       838.631
    GC Fragmentation (%)                                            2.083
    GC Heap Size (MB)                                          1,367.444
    Gen 0 GC Budget (MB)                                              94
    Gen 0 GC Count (Count / 1 sec)                                     0
    Gen 0 Size (B)                                            27,663,000
    Gen 1 GC Count (Count / 1 sec)                                     0
    Gen 1 Size (B)                                                     0
    Gen 2 GC Count (Count / 1 sec)                                     0
    Gen 2 Size (B)                                                     0
    IL Bytes Jitted (B)                                          517,522
    LOH Size (B)                                              8.0962e+08
    Monitor Lock Contention Count (Count / 1 sec)                      0
    Number of Active Timers                                            0
    Number of Assemblies Loaded                                      126
    Number of Methods Jitted                                       5,449
    POH (Pinned Object Heap) Size (B)                            168,376
    ThreadPool Completed Work Item Count (Count / 1 sec)             100
    ThreadPool Queue Length                                            0
    ThreadPool Thread Count                                           13
    Time paused by GC (ms / 1 sec)                                     0
    Time spent in JIT (ms / 1 sec)                                 4.699
    Working Set (MB)                                          1,468.977
```

Figure 4.14 – The process' real-time memory usage

The next step is to collect the correct data for memory analysis. A frequently used method for diagnosing such issues involves creating a memory dump on Windows or a core dump on Linux. This will capture the state of the application's memory. The dotnet-dump tool is commonly employed to produce these dumps for .NET applications. It offers a solution for gathering and examining dumps across Windows, Linux, and macOS platforms without requiring a native debugger. With dotnet-dump, you can execute SOS commands to delve into issues related to crashes and the garbage collector.

To install this tool, you must run the following command in a terminal window:

```
dotnet tool install --global dotnet-dump
```

Once installed, you can run the following command against your process to generate a dump:

```
dotnet-dump collect -p 31876
```

Now that you have generated a core dump, use the dotnet-dump tool to analyze the dump:

```
dotnet-dump analyze dump_20240409_124020.dmp
```

Note that dump_20240409_124020.dmp should be replaced with the name of the generated dump. This command opens a console with a prompt where you can enter commands. The first thing you want to look at is the overall state of the managed heap using this command:

```
dumpheap -stat
```

This outputs a table with four columns:

- **MT**: A method table containing the address for the methods being called
- **Count**: The number of instances
- **TotalSize**: The amount of memory being used
- **Class name**: The class containing the method referenced by the method in the MT column

The table is sorted by TotalSize in ascending order, so scrolling to the end of the table will reveal when larger objects begin to exist and what their type is. In *Figure 4.15*, we can clearly see that the System.String class is using an excessive amount of memory. The next step is to confirm that this class type might be used more than it should be and to confirm the method that might be misusing it.

```
7ff8a4c2dae8        2       18,528 System.Text.Encodings.Web.OptimizedInboxTextEncoder
7ff8a4c57eb0      797       19,128 System.Diagnostics.Tracing.IncrementingPollingCounterPa
7ff8a4b02800      246       23,616 Microsoft.Extensions.DependencyInjection.ServiceLookup.
7ff8a4edea90       48       25,728 System.Collections.Concurrent.ConcurrentQueueSegment<Sy
7ff8a4edf958       48       25,728 System.Collections.Concurrent.ConcurrentQueueSegment<Mi
7ff8a457fab0      113       26,200 System.String[]
7ff8a4b009a0      418       26,752 System.Collections.Concurrent.ConcurrentDictionary<Micr
p.ServiceCallSite>+Node
7ff8a47a93b8      106       30,528 System.GCMemoryInfoData
7ff8a461b2b0      795       32,264 System.RuntimeType[]
7ff8a461d898      313       32,552 System.Reflection.RuntimeConstructorInfo
7ff8a4edf098       48       38,016 System.Collections.Concurrent.ConcurrentQueueSegment<Mi
7ff8a4641198      766       38,112 System.Reflection.ParameterInfo[]
7ff8a4615bd8       10       38,736 System.Diagnostics.Tracing.EventSource+EventMetadata[]
7ff8a4425fa8    1,691       40,584 System.Object
7ff8a46bfa50      381       42,272 System.Diagnostics.Tracing.DiagnosticCounter[]
7ff8a45f6258      286       45,760 System.RuntimeType+RuntimeTypeCache
7ff8a4641070      769       61,520 System.Signature
7ff8a4c27518    2,771       66,504 System.Diagnostics.Tracing.PollingPayloadType
7ff8a442a318    1,665       66,600 System.RuntimeType
7ff8a4c57d78      797       70,136 System.Diagnostics.Tracing.IncrementingCounterPayload
7ff8a442c4d8      198       88,080 System.Object[]
7ff8a44d9df8      151      113,408 System.Int32[]
7ff8a4610f58    1,402      145,808 System.Reflection.RuntimeMethodInfo
7ff8a4611fe0    1,586      152,256 System.Reflection.RuntimeParameterInfo
7ff8a45f2c78      681      240,694 System.Byte[]
7ff8a4c273e0    3,248      337,792 System.Diagnostics.Tracing.CounterPayload
025a15114b90    1,269      364,696 Free
7ff8a44dec08   10,903 2,435,399,820 System.String
Total 41,586 objects, 2,438,315,096 bytes
```

Figure 4.15 – The string class type using excess amounts of memory

Now, we can use the `dumpheap —mt 7ff8a44dec08` command to analyze the string type's usage. *Figure 4.16* shows the result of this command, where all the string instance calls are logged in a table and then a final usage table is presented.

```
025ade200160        7ff8a44dec08        2,097,174
025adf800048        7ff8a44dec08        2,097,174
025adfa00080        7ff8a44dec08        2,097,174
025adfc000b8        7ff8a44dec08        2,097,174
025adfe000f0        7ff8a44dec08        2,097,174
025ae0000128        7ff8a44dec08        2,097,174
025ae0200160        7ff8a44dec08        2,097,174
025ae1800048        7ff8a44dec08        2,097,174
025ae1a00080        7ff8a44dec08        2,097,174
025ae1c000b8        7ff8a44dec08        2,097,174
025ae1e000f0        7ff8a44dec08        2,097,174
025ae2000128        7ff8a44dec08        2,097,174
025ae2200160        7ff8a44dec08        2,097,174
025ae2400198        7ff8a44dec08        2,097,174

Statistics:
          MT   Count      TotalSize Class Name
7ff8a44dec08   10,903 2,435,399,820 System.String
Total 10,903 objects, 2,435,399,820 bytes
```

Figure 4.16 – The string class type statistics and method misusing it

Using one of the address references of a large string, we can run the `gcroot` command to show a stack trace of the instance's usage in the application. This is shown in *Figure 4.17*.

```
> gcroot 025ae2400198
HandleTable:
    0000025a16a811f8 (strong handle)
          -> 025a21000028    System.Object[]
          -> 025a5001b1d0    System.Collections.Generic.List<System.String>
          -> 025a5001be38    System.String[]
          -> 025ae2400198    System.String

Thread d770:
    c9832ff7f0 7ff8f94ee07a System.Diagnostics.Tracing.CounterGroup.PollForValues() [/_/src/libraries/System.Private.CoreLib/src/System/Diagnostics/Tracing/CounterGroup.cs @ 305]
        rsi:
          -> 025a21000028    System.Object[]
          -> 025a5001b1d0    System.Collections.Generic.List<System.String>
          -> 025a5001be38    System.String[]
          -> 025ae2400198    System.String

Thread 9528:
    c9834bf820 7ff8f93f8f36 System.Threading.PortableThreadPool+GateThread.GateThreadStart() [/_/src/libraries/System.Private.CoreLib/src/System/Threading/PortableThreadPool.GateTh
read.cs @ 68]
        r14:
          -> 025a21000028    System.Object[]
          -> 025a5001b1d0    System.Collections.Generic.List<System.String>
          -> 025a5001be38    System.String[]
          -> 025ae2400198    System.String

        rbp-90: 000000c9834bf8c0
          -> 025a21000028    System.Object[]
          -> 025a5001b1d0    System.Collections.Generic.List<System.String>
          -> 025a5001be38    System.String[]
          -> 025ae2400198    System.String
```

Figure 4.17 – The stack trace of the string instance's usage

This result directly correlates the string instance and a list of string objects. This means that we can now isolate the method being investigated during the application's runtime and look for all instances of a list using string objects. This investigation will help us narrow down the more minute parts of the method that may prove problematic. You can continue dumping out objects to see that most string objects follow a similar pattern. At this point, the investigation has provided sufficient information to identify the root cause in your code.

Now that we know how to review our application's memory usage, let's focus on some best development patterns that can help us develop more bulletproof applications.

Best practices for avoiding memory leaks

We have been looking at this topic for a while now. Some of the strategies that will be outlined here will have been mentioned before. We must deeply understand the implications of some coding practices and know why we do and don't do certain things. We also need to be comfortable with certain coding patterns that help to reduce the risks of memory leaks. First, we will review the practice of using the `using` keyword.

Using the using keyword with IDisposable objects

The `using` statement ensures that the `Dispose` method is called automatically, which helps free resources even if an exception occurs. Recall that the `using` statement in C# is a syntactic construct that ensures the proper use and disposal of resources that implement the `IDisposable` interface. This is crucial for managing unmanaged resources such as file handles, database connections, and network streams. When an object is used within a `using` block, its `Dispose` method is automatically called at the end of the block, even if an exception occurs within the block.

Common scenarios for using the using statement include file, database, and HTTP request scenarios:

- **Working with files**: When dealing with file operations, it is essential to ensure that file handles are closed after the operations are completed to avoid locking issues and memory leaks.

- **Working with database connections**: Database connections are another common area where resources must be managed carefully to prevent connection pool exhaustion and memory leaks.

- **Working with HTTP calls**: When making HTTP calls, the response and stream must be disposed of to release network resources.

This ensures that the garbage cleaner cleans up the resource and that we don't end up with a memory leak each time an object of this type is used. Additional benefits of the using statement include the following:

- **Simplified code**: Automatically handles the cleanup of resources, reducing boilerplate code for resource management

- **Enhanced safety**: Makes sure that resources are appropriately released, which helps to maintain application stability and performance

- **Improved readability**: Makes the resource management code explicit, showing where resources are scoped and released

Sometimes, we need to implement our own IDisposable methods and code. In this case, we must use the IDisposable and Finalizer patterns. We will investigate these next.

Implementing IDisposable and Finalizer patterns

The **Dispose/Finalize pattern** in .NET is a design approach for cleaning up unmanaged resources. This pattern is crucial for preventing resource leaks and ensuring that resources are released in a timely manner, particularly in environments where GC alone isn't sufficient for resource management.

The pattern revolves around two key methods:

- Dispose(): Part of the IDisposable interface, this method explicitly releases managed and unmanaged resources. Consumers of the object should call it when they are finished using it.

- Finalizer: This method acts as a destructor and is automatically called by the garbage collector when the object is being collected. It is typically used to release unmanaged resources if Dispose was not called.

The following is a code snippet that shows a typical implementation of a custom class that implements the IDisposable interface. In this example, we are implementing a unit-of-work wrapper around database objects used in a .NET Core application. A unit of work represents a body of work that should be completed. This pattern is typically implemented in data-driven applications that must implement data transactions to reduce data loss between operations.

We use an Entity Framework Core database context to connect to a database and carry out its operations. This context or connection should be used for as long as an operation takes and must be closed and disposed of afterward. While this is the default behavior of the database context, it becomes tricky when we create an abstraction layer, or repository and unit of work layer, on top of the database context.

Let us simulate using these abstraction layers in building a leave management system. We will start with an entity called `LeaveType`, which is defined in the ensuing code snippet:

```
namespace LeaveManagement.Data;

public class LeaveType : BaseEntity
{
    public string Name { get; set; }
    public int DefaultDays { get; set; }
}
```

We also have an Entity Framework database context, which enlists the data model and identifies it as an accessible `DbSet`:

```
namespace LeaveManagement.Data
{
    public class ApplicationDbContext : DbContext
    {
        public ApplicationDbContext(DbContextOptions
        <ApplicationDbContext> options)
            : base(options)
        {
        }

        public DbSet<LeaveType> LeaveTypes { get; set; }
    }
}
```

We also have a generic repository interface, as defined in the following:

```
using LeaveManagement.Data;
namespace LeaveManagement.Application.Contracts
{
    public interface IGenericRepository<T> where T : class
    {
        Task<T> GetAsync(int? id);
        Task<List<T>> GetAllAsync();
        Task<T> AddAsync(T entity);
        Task AddRangeAsync(List<T> entities);
        Task<bool> Exists(int id);
```

```
        Task DeleteAsync(int id);
        Task UpdateAsync(T entity);
    }
}
```

This IGenericRepository <T> is implemented by GenericRepository:

```
namespace LeaveManagement.Application.Repositories
{
    public class GenericRepository<T> : IGenericRepository<T>
    where T : class
    {
        private readonly ApplicationDbContext context;

        public GenericRepository(ApplicationDbContext context)
        {
            this.context = context;
        }

        public async Task<T> AddAsync(T entity)
        {
            await context.AddAsync(entity);
            await context.SaveChangesAsync();
            return entity;
        }

        public async Task AddRangeAsync(List<T> entities)
        {
            await context.AddRangeAsync(entities);
            await context.SaveChangesAsync();
        }

        public async Task DeleteAsync(int id)
        {
            var entity = await GetAsync(id);
            context.Set<T>().Remove(entity);
            await context.SaveChangesAsync();
        }

        public async Task<bool> Exists(int id)
        {
            var entity = await GetAsync(id);
            return entity != null;
        }
```

```
public async Task<List<T>> GetAllAsync()
{
    return await context.Set<T>().ToListAsync();
}

public async Task<T> GetAsync(int? id)
{
    if (id == null)
    {
        return null;
    }
    return await context.Set<T>().FindAsync(id);
}

public async Task UpdateAsync(T entity)
{
    context.Update(entity);
    await context.SaveChangesAsync();
}
    }
}
```

Now that the repository layer is in place, we need to add it to a unit of work to ensure that all operations performed in instances of GenericRepository operate as a unit. Now we will have one implementation of a SaveChanges() operation in a method called Save(). First, we will have an interface that outlines the properties to be implemented in the derived class:

```
public interface IUnitOfWork : IDisposable
{
    IGenericRepository<LeaveType> LeaveTypes { get; }
    Task Save();
}
```

The interface inherits from IDisposable, so the derived class will also inherit it. The derived class is as follows:

```
public class UnitOfWork : IUnitOfWork
{
    private readonly ApplicationDbContext _context;
    //... properties omitted for brevity

    public UnitOfWork(ApplicationDbContext context)
    {
```

```
            _context = context;
        }
        public void Dispose()
        {
            Dispose(true);
            GC.SuppressFinalize(this);
        }

        private void Dispose(bool dispose)
        {
            if(dispose)
            {
                _context.Dispose();
            }
        }

        public async Task Save()
        {
            await _context.SaveChangesAsync();
        }
    }
```

This class uses DI to instantiate an object in the database context. We then have several additional methods that are required based on the pattern:

- `Dispose()`: This method is explicitly called to free resources. It calls the protected `Dispose(bool dispose)` method with `dispose` set to `true`.

- `Dispose(bool dispose)`: This method performs the actual work of releasing resources. If `dispose` is `true`, it disposes of managed resources (such as other `IDisposable` objects). Regardless of the disposing parameter, it should always free unmanaged resources and ensure that this operation is only performed once by checking the `disposed` field.

- `Finalizer`: If the user forgets to call `Dispose()`, the garbage collector calls the finalizer (destructor). It executes the same cleanup logic but specifies that managed resources should not be freed, as the finalizer is only concerned with unmanaged resources.

- `GC.SuppressFinalize(this)`: This call prevents the finalizer from running if `Dispose()` has already been called, optimizing the GC process by avoiding unnecessary finalization.

Some considerations around using this pattern are as follows:

- Always implement the `Dispose` pattern when dealing with unmanaged resources.

- Encourage consumers of your class to use the `using` statement to ensure that resources are automatically cleaned up.

- Consider thread safety when implementing the `Dispose` pattern if the object is accessed from multiple threads.

- Every type with a finalizer should implement `IDisposable` after calling `Dispose` to ensure that the object is unusable.

- When you are finished using an `IDisposable` type, call the `Dispose()` method.

- Allow multiple calls to the `Dispose()` method without raising errors.

- Leverage the `GC.SuppressFinalize()` method to suppress subsequent finalizer calls within the `Dispose()` method.

- Do not create value types that are disposable.

- Throwing exceptions within the disposal methods is not recommended.

The `Dispose/Finalize` pattern effectively manages managed and unmanaged resources in .NET. In this example, I mention the concept of DI, whereby the database context is used. We will now explore how this pattern helps prevent memory leaks.

Using Dependency Injection

DI is a design pattern used in software development to manage dependencies between objects. By injecting dependencies rather than having objects create them internally, DI promotes loose coupling and greater modularity, which can have several benefits, including reducing the risk of memory leaks. Here's how DI can help prevent memory leaks:

- **Managed life cycle**: DI frameworks typically manage the life cycles of dependencies. This means that the framework creates and destroys objects, ensuring that the memory allocated for dependencies is well managed. By controlling when an object is disposed of, the framework can help prevent the memory that it occupies from being leaked because it ensures that cleanup code, such as destructors or disposal methods, is called.

- **Consistent object creation and destruction**: Since the DI framework handles object creation and destruction, it minimizes the risks associated with manual memory management, whereby developers might forget to free allocated memory. The consistency in how objects are created and destroyed helps in avoiding leaks.

- **Reducing object scope**: DI can help reduce the scope of objects to where they are needed. By injecting objects only into components where they are required, DI minimizes the existence of unused objects floating around in the application's memory space, thus reducing the risk of memory leaks.

- **Improved testability and maintenance**: With DI, components are easier to test and maintain. Test environments can use mock or stub implementations of interfaces that can be injected into components. This separation of concerns ensures that memory management testing can be done more rigorously and in isolation, helping to catch memory leaks early in development.

- **Encouraging the single responsibility principle**: By decoupling object creation from business logic, DI encourages adherence to the **single responsibility principle**, which states that a class or module should have just one reason to change. This leads to better-organized code that is easier to debug and less likely to have unforeseen memory leaks due to overcomplicated object interactions and life cycle management mishaps.

- **Proper resource management**: In many cases, dependencies involve resources that need explicit release, such as file handles, network connections, or database connections. DI frameworks often support the automatic management of such resources, ensuring they are released when they are no longer needed.

While DI doesn't automatically prevent memory leaks, it facilitates better design patterns and practices that reduce the likelihood of their occurrence. By using DI, developers can focus more on the business logic and less on the intricacies of memory and resource management, relying on the DI framework to handle much of this responsibility.

The ASP.NET Core offers robust support for DI with its integrated DI container, where you can register your services. Services can be registered in three modes: **singleton, transient**, or **scoped**. The DI container manages its disposal appropriately for services that implement the `IDisposable` interface. Consequently, while Transient and Scoped service instances are disposed of at the end of a request or scope, respectively, Singleton service instances are disposed of only when the application shuts down.

Consider the following example where DI is used to manage the service life cycle that interacts with an external API. Here, we will use the `HttpClient` class, which represents an unmanaged resource that must be disposed of properly to avoid a memory leak:

```
public interface IWeatherService
{
    Task<string> GetWeatherAsync(string city);
}
public class OpenWeatherMapService : IWeatherService
{
    private readonly HttpClient _httpClient;
    public OpenWeatherMapService(HttpClient httpClient)
    {
        _httpClient = httpClient;
    }

    public async Task<string> GetWeatherAsync(string city)
    {
```

```
        string requestUrl = $"https://api.openweathermap.org/data/2.5/
        weather?q={city}";
        HttpResponseMessage response =
        await _httpClient.GetAsync(requestUrl);
        response.EnsureSuccessStatusCode();
        return await response.Content.ReadAsStringAsync();
    }
}
```

Here, we have an interface and implementing class where we make an API call to the external resource. To facilitate this operation, we inject an instance of an `HttpClient` object. This object is best used only for the scope of the request, which means that we should use a `using` statement for the block of code that it is being called in, or that we should use an appropriate method of DI.

To register this service, we have several options:

- **Scoped**: Services are created once per client request (per scope) with this option. A new instance is created for each request but shared across different components within that request. This is useful for services that maintain state within a request but should not be shared across different requests. This is particularly important in web applications where each request might serve different users.

  A database connection (such as an entity framework core database context) would be a good candidate for this lifetime type.

- **Transient**: Services are created each time they are requested from the service container with this option. This means that if multiple components depend on a transient service, each receives a separate instance. These services are best for lightweight, stateless, independent services that do not need to maintain any state between calls. Since a new instance is created on every request, there is no need to worry about thread safety concerning class fields. A logging service where there is no need to retain data in between requests would be a good candidate for this type of lifetime.

- **Singleton**: Services are created once and shared throughout the application's lifetime with this option. Once instantiated, the same service instance is used on every request. Singleton services are appropriate for stateless services or those that have a read-only state, as they are shared across multiple requests and threads. Using singletons for stateful, shared services can lead to concurrency issues. A configuration service that reads values once from a file and stores them would be a good candidate for this lifetime since the configuration file's data does not change and it is safe to be used by concurrent requests.

- **AddHttpClient**: This option registers the client with a **transient** lifetime. Each time a client is injected, it receives a new instance of `HttpClient` that `IHttpClientFactory` manages. The `IHttpClientFactory` handles the underlying network handlers' efficient pooling and life cycle management.

We registered the service using the `AddHttpClient` method and allowed the DI's default dependency management method (transient) to implement it. This setup ensures that each service can use the `HttpClient` instances as needed without worrying about the complexities of managing connection lifetimes while avoiding potential problems such as memory leaks or socket exhaustion. This approach balances performance, scalability, and reliability in web applications that rely heavily on outbound HTTP calls.

We can do that by adding the following code to the `Program.cs` file:

```
builder.Services.AddHttpClient<IWeatherService,
OpenWeatherMapService>();
```

Now, we can safely inject this `IWeatherService` anywhere in our code and not need to worry about its disposal and memory leak concerns outside of poorly written code inside the implementation of the method:

```
public class WeatherController : ControllerBase
{
    private readonly IWeatherService _weatherService;

    public WeatherController(IWeatherService weatherService)
    {
        _weatherService = weatherService;
    }

    [HttpGet("{city}")]
    public async Task<IActionResult> GetWeather(string city)
    {
        try
        {
            var weatherData =
            await _weatherService.GetWeatherAsync(city);
            return Ok(weatherData);
        }
        catch (HttpRequestException ex)
        {
            return BadRequest("Error fetching weather data");
        }
    }
}
```

Each time this endpoint is called, a new service instance will be instantiated and destroyed, whether the request was successful or not.

Now that we have examined some practical methods for managing and avoiding memory leaks, let's summarize this chapter.

Summary

This chapter explored various techniques and tools to detect memory leaks in applications. It emphasized the importance of recognizing the signs of memory leaks, such as gradually increasing memory usage over time, application slowdowns, or crashes due to exhausted resources.

We also reviewed memory analysis, using .NET-specific tools and methodologies to examine how .NET applications utilize memory. We covered using the memory usage profilers and diagnostic tools in Visual Studio. These tools give us real-time feedback on how an application is performing during debugging and are extremely useful for early memory leak detection. We then reviewed how to use extensions in the .NET CLI to analyze data dumps and memory usage. This is a less visually appealing option, but it is useful for non-Windows-based machines.

Finally, we reviewed a set of best practices designed to prevent memory leaks before they happen:

- Proper management of disposable resources using patterns such as `using` and the `IDisposable` interface
- Avoiding global static variables that can prevent GC
- Regular code reviews and testing during the development cycle to catch memory leaks early
- Stress testing your application for early detection of potential memory leaks under load.
- Adopting a coding style that emphasizes cleanliness and simplicity reduces the chance of unintentionally holding onto memory

We should strive to refine our skills in building efficient, leak-free applications. While tools and techniques are essential, awareness, and adherence to best practices are equally crucial in preventing memory leaks. In the next chapter, we'll cover some advanced memory management techniques.

5

Advanced Memory Management Techniques

Memory management is a cornerstone in .NET programming, underpinning application performance and stability. As developers, we often encounter complex scenarios that challenge our understanding of how memory is allocated, managed, and optimized, especially in modern applications that leverage concurrency and asynchrony.

This chapter delves into advanced memory management techniques within the .NET ecosystem, focusing on three critical areas:

- Concurrent memory management
- Memory usage in multi-threaded applications
- Memory management in asynchronous code

.NET provides several mechanisms to handle memory safely when multiple threads operate in parallel, which is increasingly common in today's software landscape. We will examine the role of garbage collection in a concurrent environment, including the optimizations that .NET implements to minimize pause times and enhance throughput.

We will also explore memory usage in multi-threaded applications. Multi-threading allows an application to perform multiple operations simultaneously, improving performance and responsiveness. However, this can lead to complex memory management challenges such as race conditions, deadlocks, and resource contention.

Finally, we will address memory management in asynchronous code – a vital feature of modern .NET applications, especially in web services and I/O-bound operations. Asynchronous programming in .NET helps create scalable applications and introduces unique memory management challenges. We will explore how the .NET runtime handles memory allocation and deallocation in asynchronous operations, the impact of closures and captured variables on the memory life cycle, and techniques for avoiding common pitfalls such as memory leaks and excessive memory allocation.

We will go over the following main topics in this chapter:

- Concurrent memory management

- Memory usage in multi-threaded applications

- Memory management in asynchronous code

This chapter's practical examples, code snippets, and real-world scenarios illustrate the concepts discussed. By the end of this chapter, you will have a deep understanding of how advanced memory management techniques can be applied in .NET programming to build efficient, robust, and scalable applications.

Technical requirements

For this chapter, you will require the following:

- Visual Studio 2022 Studio (https://visualstudio.microsoft.com/vs/community/)

- Visual Studio Code (https://code.visualstudio.com/)

- .NET 8 SDK (https://dotnet.microsoft.com/en-us/download/visual-studio-sdks)

Concurrent memory management

Concurrent programming, often called multithreading or parallel programming, allows multiple processes to execute simultaneously, enhancing the performance of applications, especially on multi-core processors. In .NET, concurrent programming is supported through various constructs, such as threads, thread pools, tasks, and asynchronous programming models. However, while concurrency can dramatically improve the efficiency and responsiveness of applications, it introduces challenges and quirks requiring specialized techniques to manage effectively.

Concurrency in .NET programming refers to an application's ability to manage multiple tasks simultaneously, allowing it to perform more than one operation simultaneously. This is particularly important in modern computing environments where applications must be responsive, efficient, and capable of handling multiple operations or large volumes of data without delay.

Some of the quirks and challenges of concurrent programming in .NET are as follows:

- **Race conditions**: A race condition occurs when two or more threads attempt to modify a shared resource simultaneously, leading to unpredictable outcomes. This is a common issue in concurrent programming, where the sequence of execution matters.

- **Deadlocks**: A deadlock happens when two or more threads each hold a resource and wait for the other to release another resource, creating a cycle of dependencies that prevents any of them from proceeding.

- **Starvation**: Starvation occurs when a thread is perpetually denied access to resources it needs because other threads are monopolizing them.

- **Thread safety**: Not all code is thread-safe by default, meaning it can safely be called from multiple threads simultaneously. Non-thread-safe code can lead to data corruption when accessed by multiple threads.

- **Context switching overhead**: Every time the CPU switches from executing one thread to another, a context switch overhead is involved, which can reduce the efficiency gains from parallelism if not managed correctly.

To address these quirks and ensure successful concurrent programming in .NET, developers must employ specialized techniques:

- **Locking mechanisms**: Utilizing locks (lock keyword in C#) is a common way to prevent race conditions by ensuring that only one thread can access a resource at a time. However, excessive use of locks can lead to deadlocks and reduce performance.

- **Immutable objects**: Immutable objects cannot be modified after their creation. Using immutable objects can simplify concurrent programming because there is no need to worry about one thread modifying the state that another thread is depending on.

- **Concurrent collections**: .NET provides thread-safe versions of common collection types, which manage synchronization internally and reduce the developer's burden while mitigating common concurrency issues.

- **Task Parallel Library (TPL) and Parallel LINQ (PLINQ)**: TPL simplifies writing concurrent and asynchronous code by abstracting much of the complexity involved in directly managing threads. PLINQ can also efficiently execute data queries in parallel, automatically handling thread management and synchronization.

- **Asynchronous Programming Model (APM)**: Asynchronous methods help prevent thread blocking and improve application responsiveness, particularly in I/O-bound scenarios. The async-await pattern in C# is a powerful feature for this purpose.

Effective concurrent programming in .NET requires a deep understanding of potential pitfalls and the specialized techniques available to avoid them. By leveraging .NET's powerful frameworks and libraries and following best practices for thread safety and memory management, developers can create efficient, robust, and scalable multi-threaded applications.

Next, we'll review threads and thread pooling.

Threads and thread pooling

Threading is a fundamental concept that allows multiple operations to run concurrently within a single application. This capability enhances applications' responsiveness and presents unique memory usage and management challenges.

Using threads affects memory usage in several ways:

- **Stack allocation**: Each thread in C# has its stack, typically consuming around 1 MB of memory. This stack stores local variables, method call histories, and control flow for the thread. Due to these stack allocations, creating many threads can quickly increase an application's memory footprint.

- **Heap allocation**: Although threads share data stored on the heap, concurrent operations may lead to increased memory allocation on the heap as each thread might create its objects to avoid sharing state and minimize synchronization issues.

- **Synchronization overhead**: Managing thread safety through locks, mutexes, and other synchronization mechanisms can add overhead in performance and memory as these mechanisms often involve additional data structures.

.NET provides a rich threading model. Threads are the most fundamental unit of execution in an operating system, and .NET allows developers to create and manage threads directly using the `System.Threading.Thread` class. Threads can perform tasks in parallel, improving the application's throughput and responsiveness. Here's an example of creating a thread in C#:

```
using System;
using System.Threading;
public class Example
{
    public void ExecuteThread()
    {
        Thread thread = new Thread(new ThreadStart(LongRunningTask));
        thread.Start();
    }

    private void LongRunningTask()
    {
        // simulate work on the thread.
        Thread.Sleep(10000);
        Console.WriteLine("Task completed");
    }
}
```

Creating and managing individual threads can be helpful for long-running operations. However, each thread has its overhead, including memory allocated for stack space, which typically ranges from 1 MB per thread in .NET applications. By manually managing threads, developers have the flexibility to control thread behavior but must also handle the associated memory costs. This leads us to pooling threads for more efficiency.

The `ThreadPool` class is a collection of worker threads managed by the **Common Language Runtime (CLR)** that can perform short tasks without the overhead of creating and destroying threads. It is more efficient for short-lived functions that need to be executed frequently.

The `ThreadPool` class efficiently manages a collection of threads that can be reused for multiple tasks. Unlike creating individual threads, using `ThreadPool` helps reduce the memory overhead because it limits the number of idle threads and reuses threads for numerous tasks. This is crucial for high-performance applications that need to handle many short-lived tasks. Thread pooling helps with memory management in the following ways:

- **Reduced overhead**: By reusing existing threads rather than creating new ones, `ThreadPool` reduces the memory overhead associated with thread creation (for example, stack space). Each new thread in .NET usually requires a significant amount of memory for its stack, and creating many threads can lead to high memory consumption.

- **Scalability**: The `ThreadPool` class automatically adjusts the number of threads based on the workload, which helps optimize memory usage for the application's requirements. This prevents the system from becoming overwhelmed by too many threads, which can consume excessive memory and degrade overall system performance.

- **Efficiency**: Thread pooling allows an application to handle multiple operations without thread creation and destruction delay, reducing the CPU cycles and memory allocation/deallocation overhead. This can be particularly important in services or applications with critical response time and throughput.

Here's an example of using thread pooling to simulate a web server handling multiple concurrent requests:

```
using System.Threading;
public class WebServer
{
    public void HandleRequest()
    {
        Console.WriteLine("Server ready to handle requests.");
        for (int i = 0; i < 10; i++)
        {
            int requestId = i;
            ThreadPool.QueueUserWorkItem(_ =>
            ProcessRequest(requestId));
        }
    }

    private void ProcessRequest(int requestId)
    {
        Console.WriteLine($"Processing request {requestId} on thread
        {Thread.CurrentThread.ManagedThreadId}");
```

```
        // Simulate processing time
        Thread.Sleep(2000);
        Console.WriteLine($"Request {requestId} completed.");
    }
}

public class Program
{
    public static void Main()
    {
        WebServer server = new WebServer();
        server.HandleRequest();
        Console.WriteLine("Hit Enter to exit.");
        Console.ReadLine();
    }
}
```

In this example, `ThreadPool` efficiently handles multiple requests, ensuring the application uses memory and processing resources efficiently without creating and destroying threads for each request. This is crucial for maintaining performance and resource usage under control in high-demand environments.

Managing access to shared resources across multiple threads is crucial in concurrent programming to ensure data integrity and system stability. Lock mechanisms are pivotal in controlling how multiple threads interact with shared memory. Considering this, we will review lock mechanisms next.

Lock mechanisms

.NET provides a robust environment for developing multithreaded applications. Before diving into lock mechanisms, it is essential to understand a few basic concepts in .NET concurrency:

- **Threads**: The smallest unit of processing that the operating system can schedule
- **Thread safety**: When a method or class ensures consistent behavior when executed by multiple threads simultaneously
- **Shared resources**: Data accessed by multiple threads, such as static variables or instance fields

Locking is a technique that's used to synchronize access to resources by preventing multiple threads from reading and writing to shared data concurrently. .NET offers several locking constructs:

- **The lock keyword**: The simplest way to protect a block of code is to lock it so that only one thread can execute it at a time.
- **The Monitor class**: This class provides more control over the synchronization process. In addition to basic locking, it allows setting timeouts for acquiring a lock.

- **Mutex**: This is like `Monitor` but can also be used across different processes running on the same machine.

- **Semaphore and SemaphoreSlim**: These limit the number of threads that can access a particular resource or pool of resources concurrently. They are the asynchronous alternative to the `lock` keyword.

Locking mechanisms in .NET manage thread access and enforce a memory barrier, ensuring changes in one thread's local cache are visible to other threads. Without proper locking, threads might see stale data that hasn't been updated in their local caches. Proper lock use is essential to avoid deadlocks and help prevent resource leaks by ensuring that resources are released correctly after use.

On the flip side, knowing that the **garbage collector** (**GC**) can reclaim memory used by objects that are no longer accessible, developers must ensure that locks are released. Objects are disposed of properly to aid efficient memory management. To do this, we can do the following:

- **Minimize locked time**: Keep the code inside `lock` statements to a minimum to reduce the waiting time for other threads

- **Avoid nested locks**: Nested locks can increase the complexity and risk of deadlocks

- **Use the correct locking construct**: Based on the scenario, choose the appropriate locking mechanism (for example, `Mutex` for inter-process locks and `Semaphore` for limiting concurrent access)

- **Monitor performance**: Use profiling tools to monitor the performance implications of locks in your application as excessive locking can lead to performance bottlenecks

Here, `lock` ensures that a block of code runs to completion without interruptions by other threads that might also want to execute the same code block. This is important for preventing race conditions where multiple threads might modify shared data simultaneously. Consider a scenario where we have a bank account balance that numerous threads will attempt to modify by making deposits. We can use the `lock` keyword to ensure the balance update operations are thread-safe:

```
using System;
using System.Threading;

public class BankAccount
{
    private object balanceLock = new object();
    public int Balance { get; private set; }

    // Constructor to initialize the bank account with a balance
    public BankAccount(int startingBalance)
    {
        Balance = startingBalance;
```

```
        }

        // Method to deposit money into the account
        public void Deposit(int amount)
        {
            // Locking to ensure only one thread can enter this code
            // block at a time
            lock (balanceLock)
            {
                Console.WriteLine($"Thread {Thread.CurrentThread.
                ManagedThreadId} entering deposit.");
                int initialBalance = Balance;
                Balance += amount;
                Console.WriteLine($"Deposited {amount}. Initial balance
                was {initialBalance}. New balance is {Balance}.");
            }
        }
    }

class Program
{
    static void Main(string[] args)
    {
        BankAccount account = new BankAccount(1000);

        // Create multiple threads to simulate deposit
        Thread thread1 = new Thread(() => account.Deposit(500));
        Thread thread2 = new Thread(() => account.Deposit(300));

        thread1.Start();
        thread2.Start();

        thread1.Join();
        thread2.Join();

        Console.WriteLine($"Final balance is {account.Balance}.");
    }
}
```

The BankAccount class represents a bank account with a balance and a deposit method. It includes a balanceLock object to serve as the lock token. The deposit method uses the lock keyword to synchronize access to the balance updating code. When a thread enters this method, it acquires a lock on the balanceLock object, ensuring that no other thread can enter any code block that is locked using the same object until the lock is released.

In the main method, we create two threads that simulate simultaneous deposits into the bank account. Both threads are started and then waited for using the `Thread.Join()` method.

By using the `lock` keyword, this example ensures that even if multiple threads attempt to update the balance concurrently, the operations are performed in a thread-safe manner, preventing any inconsistency in the reported balance. This example demonstrates the practical use of lock-in managing access to shared resources in a multi-threaded environment. The output is displayed in *Figure 5.1*:

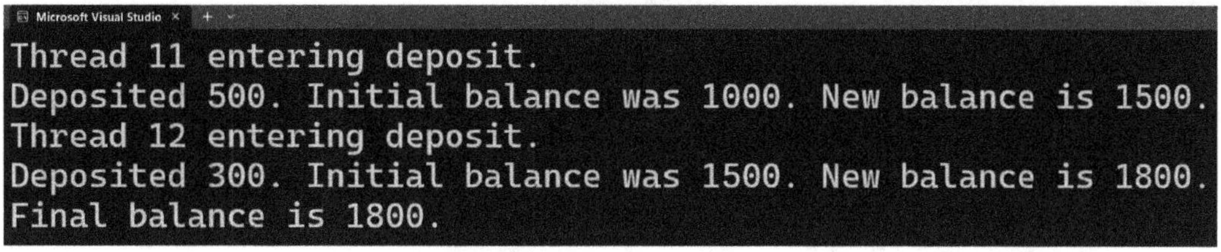

```
Thread 11 entering deposit.
Deposited 500. Initial balance was 1000. New balance is 1500.
Thread 12 entering deposit.
Deposited 300. Initial balance was 1500. New balance is 1800.
Final balance is 1800.
```

Figure 5.1 – The output of multiple threads affecting one resource

The `Monitor` class provides more flexibility than `lock`, including the ability to attempt to acquire a lock with a timeout. If we use this class to modify the preceding example, we will get code that looks like this:

```
public class BankAccount
{
    private int balance;
    private readonly object balanceLock = new object();

    // Constructor to initialize the bank account with a balance
    public BankAccount(int initialBalance)
    {
        balance = initialBalance;
    }

    // Property to safely access balance
    public int Balance
    {
        get
        {
            lock (balanceLock)
            {
                return balance;
            }
        }
        private set
        {
```

```csharp
                lock (balanceLock)
                {
                    balance = value;
                }
            }
        }


        // Method to deposit money into the account
        public void Deposit(int amount)
        {
            bool lockTaken = false;
            try
            {
                // Try to enter the lock
                Monitor.TryEnter(balanceLock, ref lockTaken);
                if (lockTaken)
                {
                    Console.WriteLine($"Thread {Thread.CurrentThread.
                    ManagedThreadId} entering deposit.");
                    int initialBalance = Balance;
                    Balance += amount;
                    Console.WriteLine($"Deposited {amount}. Initial
                    balance was {initialBalance}. New balance is
                    {Balance}.");
                }
                else
                {
                    Console.WriteLine($"Thread {Thread.CurrentThread.
                    ManagedThreadId} could not enter deposit method.");
                }
            }
            finally
            {
                // Ensure the lock is released
                if (lockTaken)
                {
                    Monitor.Exit(balanceLock);
                }
            }
        }
    }
}

class Program
```

```
{
    static void Main(string[] args)
    {
        BankAccount account = new BankAccount(1000);

        // Create multiple threads to simulate deposit
        Thread thread1 = new Thread(() => account.Deposit(500));
        Thread thread2 = new Thread(() => account.Deposit(300));

        thread1.Start();
        thread2.Start();

        thread1.Join();
        thread2.Join();

        Console.WriteLine($"Final balance is {account.Balance}.");
    }
}
```

Instead of using only the `lock` keyword, this example utilizes `Monitor.TryEnter` to attempt to acquire a lock on the `balanceLock` object. The `lockTaken` flag indicates if the lock was acquired successfully. If so, it proceeds to modify the balance; otherwise, it outputs that the thread could not enter the deposit method. This provides additional control over how long a thread waits for the lock.

The `BankAccount` class was enhanced with a getter and setter for the `Balance` property, where the value gets wrapped in a simple lock to ensure that even property accesses are thread-safe. You can think of using the lock here as syntactic sugar for `Monitor.Enter` and `Monitor.Exit`, as used in the try/finally block. *Figure 5.2* shows the output of the operation, where `thread2` was not allowed to affect the balance while `thread1` was still carrying out its operation:

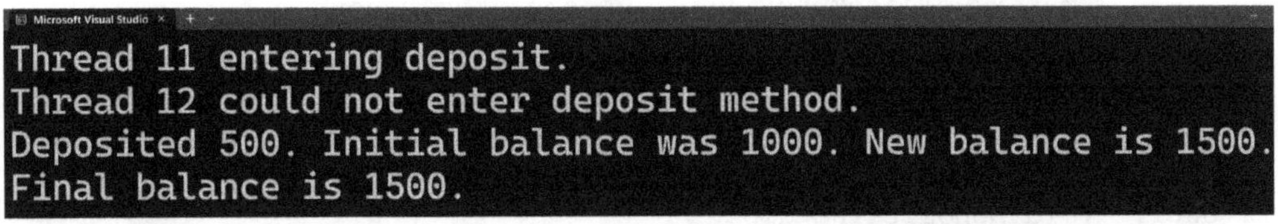

```
Thread 11 entering deposit.
Thread 12 could not enter deposit method.
Deposited 500. Initial balance was 1000. New balance is 1500.
Final balance is 1500.
```

Figure 5.2 – thread2 could not access the resource while thread1 was busy

This approach of using the `Monitor` class illustrates more advanced lock management. It offers options such as a timed wait to acquire a lock, which can be particularly useful in scenarios where waiting indefinitely is unacceptable.

Finally, let's review using `Mutex` for thread safety. A `Mutex` (or mutual exclusion) object is like `Monitor` but can be used across different processes, which makes it more versatile in specific scenarios. For our example, we'll simplify the interaction to a single application:

```csharp
public class BankAccount
{
    private int balance;
    private static Mutex mutex = new Mutex();

    // Constructor to initialize the bank account with a balance
    public BankAccount(int initialBalance)
    {
        balance = initialBalance;
    }

    // Property to safely access balance
    public int Balance
    {
        get
        {
            mutex.WaitOne(); // Acquire the mutex
            int temp = balance;
            mutex.ReleaseMutex(); // Release the mutex
            return temp;
        }
        private set
        {
            mutex.WaitOne(); // Acquire the mutex
            balance = value;
            mutex.ReleaseMutex(); // Release the mutex
        }
    }

    // Method to deposit money into the account
    public void Deposit(int amount)
    {
        mutex.WaitOne(); // Acquire the mutex before entering
                         // critical section
        try
        {
            Console.WriteLine($"Thread {Thread.CurrentThread.
            ManagedThreadId} entering deposit.");
            int initialBalance = Balance;
            Balance += amount;
```

```
            Console.WriteLine($"Deposited {amount}. Initial balance
            was {initialBalance}. New balance is {Balance}.");
        }
        finally
        {
            mutex.ReleaseMutex(); // Always release the mutex in
                                  // finally block
        }
    }
}

class Program
{
    static void Main(string[] args)
    {
        BankAccount account = new BankAccount(1000);

        // Create multiple threads to simulate deposit
        Thread thread1 = new Thread(() => account.Deposit(500));
        Thread thread2 = new Thread(() => account.Deposit(300));

        thread1.Start();
        thread2.Start();

        thread1.Join();
        thread2.Join();

        Console.WriteLine($"Final balance is {account.Balance}.");
    }
}
```

Here, we use a static Mutex object to ensure that the deposit operations are synchronized across all threads that may operate in instances of BankAccount. The Mutex object locks the critical sections where the balance is read and modified. Both the getter and setter of the Balance property and the Deposit method use Mutex to ensure exclusive access to the balance. The WaitOne method is called to request ownership of the mutex, blocking until it's acquired. After the operation, the ReleaseMutex method releases ownership, allowing other threads to proceed. We can see that both threads could conduct their operations in *Figure 5.3*:

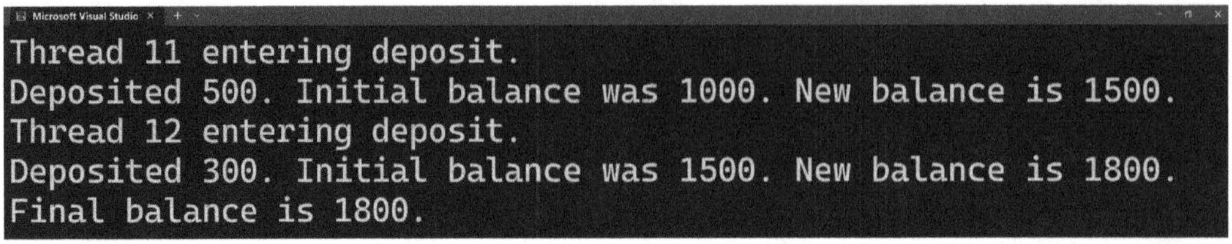

Figure 5.3 – Both threads were able to affect the resource in time

This approach is handy in scenarios where access to shared resources needs to be synchronized across different processes, although in this example, it's used within a single process for simplicity. We will review its use in a multiple-process scenario later in this chapter.

Now that we've seen how to lock down a resource to a thread, let's review TPL.

Task Parallel Library

Introduced in .NET Framework 4.0, TPL provides an abstraction over threads and simplifies parallel processing, making it easier to write concurrent and asynchronous code. The TPL uses **tasks** (`System. Threading.Tasks.Task`) to represent asynchronous operations. TPL handles the scheduling and management of these tasks efficiently using the underlying `ThreadPool`.

TPL abstracts the low-level details of thread creation and management, allowing developers to use a task-based model for concurrency. A task in .NET represents an asynchronous operation. TPL tasks are executed on the .NET thread pool, efficiently managing a background thread collection. It is an essential component of .NET, designed to make parallel programming more accessible and efficient. It provides a higher abstraction level over raw threading mechanisms, enabling developers to focus on the business logic rather than the complexities of thread management. One significant advantage of using TPL is its enhanced memory management capabilities, which include the following:

- **Reduced overhead**: Using a thread pool, TPL minimizes the overhead of creating and destroying threads. This reuse of threads can lead to significant memory savings, especially in applications that require many short-lived tasks to be executed.

- **Efficient task scheduling**: TPL includes built-in algorithms that handle task scheduling and distribution across available threads, optimizing CPU and memory usage.

- **Improved scalability**: By efficiently utilizing system resources, TPL allows applications to scale without proportional increases in memory consumption.

- **Garbage collection efficiency**: TPL's pooling approach means fewer allocations and deallocations, which reduces the GC's workload, resulting in better memory management and performance.

- **Improved resource reuse**: The memory allocated for these threads is also reused since the threads are reused rather than constantly created and destroyed. This efficient use of resources prevents frequent memory allocation and deallocation, which can strain the GC and overall system performance.

TPL promotes an asynchronous programming model, which helps manage memory more effectively. We will examine this more closely later in this chapter. The following code example shows how to create a simple task that computes a result. Notice how TPL abstracts the threading details:

```csharp
using System;
using System.Threading.Tasks;

public class Program
{
    public static void Main()
    {
        Task<int> task = Task.Run(() => ComputeResult());
        Console.WriteLine($"Result: {task.Result}");
    }

    private static int ComputeResult()
    {
        // Simulate some computation
        return new Random().Next(100);
    }
}
```

This example demonstrates the simplicity of executing a function asynchronously without manually managing the thread lifecycle.

Another critical scenario for using tasks to build robust and memory-efficient applications is task continuation. **Task continuations** are a powerful feature of TPL that allows developers to create sequences of functions that run one after the other. This feature simplifies the management of asynchronous operations, making it easier to perform subsequent actions once a task has been completed, failed, or canceled.

Continuations allow developers to pass state from one task to another in a controlled manner. This reduces the need for additional allocations to store intermediate states as each continuation has access to the results of its antecedent task. Simplified state management means fewer temporary objects and less pressure on the GC, contributing to better overall memory efficiency.

Task continuations can be set up using the ContinueWith method or by utilizing asynchronous programming patterns with await. The ContinueWith method takes a delegate that specifies what should happen when the preceding task finishes. This delegation allows for precise, logical chaining of operations without the need to nest callbacks or manually manage complex error handling and state propagation. Here's an example of setting up a continuation:

```
using System;
using System.Threading.Tasks;

public class Program
{
    public static void Main()
    {
        Task task1 = Task.Run(() => Console.WriteLine("First Task"));
        Task task2 = task1.ContinueWith(previousTask =>
        Console.WriteLine("Second Task"));

        Task.WaitAll(task2);
    }
}
```

In this example, the continuation task automatically starts after the initial task is completed and uses the result of the initial task in its execution. This pattern minimizes resource consumption by chaining tasks rather than running them concurrently.

You will notice that we also employed the WaitAll() method. This method blocks the calling thread until all the provided tasks have been completed. Several other methods either block the thread until an operation is done or return data when any task is completed.

The WaitAny() method blocks the calling thread until any provided tasks have been completed:

```
Task[] tasks = { Task1(), Task2(), Task3() };
Task.WaitAny(tasks); // unblocks when any task is finished
```

Unlike WaitAll(), it will return data when any tasks in the list have finished their work. Here, Wait blocks the calling thread until a task has been completed. This can be used with an async method that is not called asynchronously with the await keyword:

```
Task task = Task1();
task.Wait(); // blocks the thread until the task is completed
```

Here, `GetAwaiter().GetResult()` blocks the calling thread until the task has been completed and then returns the result. This is also used with a `Result` property to retrieve the result of an asynchronous method called synchronously:

```
Task<int> task = Task1();
int result = task.Result; // synchronously called the method and
returns the data.
```

The `WhenAll()` and `WhenAny()` methods are non-blocking and can be called with the `await` keyword. The `WhenAll()` method returns a completed task when all the provided tasks have been completed, while the `WhenAny()` method returns a task when any provided tasks have been completed:

```
Task[] tasks = { Task1(), Task2(), Task3() };
await Task.WhenAll(tasks); // only returns when all tasks are done
await Task.WhenAny(tasks); // returns as soon as at least one task is
done.
```

Now that we understand TPL and how tasks can be created and used in programming, let's review how resources can be locked using the `SemaphoreSlim` keyword.

SemaphoreSlim for task-based synchronization

Asynchronous methods typically use the `await` keyword, which allows the technique to yield control back to the caller and resume later. This can cause problems with `lock` because the execution context may switch, leading to situations where the lock is not correctly released or acquired by different threads, causing unpredictable behavior. Because the `lock` keyword is blocking, when a thread tries to enter a locked code section, it will block if another thread has already acquired the lock. This blocking behavior is unsuitable for asynchronous code, which aims to avoid blocking and keep the application responsive.

When using await within a lock statement, context switching can lead to deadlocks because the execution context may change, preventing the lock from being properly released or acquired.

The `SemaphoreSlim` class is a lightweight, non-blocking synchronization primitive that's suitable for asynchronous programming. It allows you to limit the number of tasks that can access a resource concurrently and provides methods designed to work with `async` and `await`. Unlike `lock`, `SemaphoreSlim` allows you to wait asynchronously for the semaphore to become available without blocking a thread, as shown in the following code snippet:

```
namespace AsyncCounterExample
{
    class Program
    {
        static async Task Main(string[] args)
        {
```

```csharp
            AsyncCounter counter = new AsyncCounter();

            // Create an array of tasks to demonstrate concurrent
            // increments
            Task[] tasks = new Task[10];
            for (int i = 0; i < tasks.Length; i++)
            {
                tasks[i] = Task.Run(async () =>
                {
                    await counter.IncrementAsync();
                    Console.WriteLine($"Count after increment:
                    {await counter.GetCountAsync()}");
                });
            }

            // Wait for all tasks to complete
            await Task.WhenAll(tasks);

            // Final count
            Console.WriteLine($"Final count:
            {await counter.GetCountAsync()}");
        }
}

public class AsyncCounter
{
    private int _count = 0;
    private SemaphoreSlim _semaphore = new SemaphoreSlim(1, 1);
    // Initial and maximum count is 1

    public async Task IncrementAsync()
    {
        await _semaphore.WaitAsync();
      // Asynchronously wait until the semaphore is available
        try
        {
            _count++;
        }
        finally
        {
            _semaphore.Release(); // Release the semaphore
        }
    }
```

```
public async Task<int> GetCountAsync()
{
    await _semaphore.WaitAsync();
    try
    {
        return _count;
    }
    finally
    {
        _semaphore.Release();
    }
}
}
}
```

Here, we define a class called AsyncCounter that encapsulates the logic for an asynchronous counter with thread-safe increment and gets operations using SemaphoreSlim. The IncrementAsync and GetCountAsync methods increment the counter in a thread-safe manner by waiting for the semaphore to be available. In the main method, we create an instance of this class; Task.Run is used to create tasks that call the IncrementAsync method of the AsyncCounter class. Then, Task. WhenAll(tasks) executes the tasks and returns the result, as shown in *Figure 5.4*:

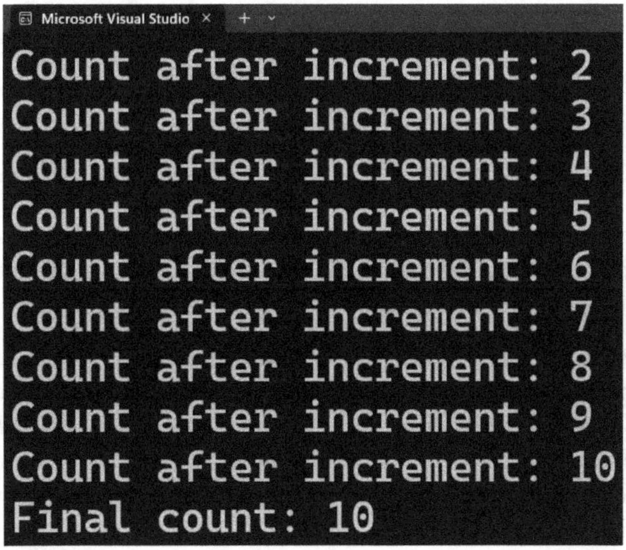

Figure 5.4 – The result of using SemaphoreSlim with asynchronous tasks

Now that we've reviewed how tasks work, we will review PLINQ, a specialized version of the standard LINQ toolset.

PLINQ

Let's begin by reviewing **LINQ**, a set of technologies introduced in .NET Framework 3.5 that extends powerful query capabilities to the syntax of C# and other .NET languages. LINQ stands for **Language Integrated Query** and enables developers to write queries directly against arrays, enumerable classes, XML documents, relational databases, and more using a consistent, readable syntax integrated into the programming language.

LINQ is primarily used for querying data. The syntax of LINQ is modeled on **Structured Query Language** (**SQL**) but is designed to work with different data sources. Here are some common uses of LINQ:

- **Querying collections**: LINQ can query any .NET collection implementing `IEnumerable<T>` or `IQueryable<T>`, allowing for filtering, ordering, and aggregation with minimal code

- **XML data**: LINQ to XML provides an in-memory XML programming interface that enables easy manipulation and querying of XML documents

- **Database access**: LINQ to SQL and Entity Framework allows you to query relational databases by translating LINQ queries into SQL queries

This simple code example of LINQ can retrieve elements from a list:

```
var numbers = new List<int> { 1, 2, 3, 4, 5, 6 };
var evenNumbers = from num in numbers
                  where num % 2 == 0
                  select num;
```

PLINQ is an extension of LINQ that allows developers to utilize the power of parallel computing within their .NET applications. PLINQ enables the execution of queries across multiple processors or cores, thus enhancing performance by distributing workloads efficiently.

PLINQ is a part of TPL and integrates seamlessly with the existing LINQ architecture. Using PLINQ allows developers to parallelize their data queries with minimal changes to their code base. The key concept behind PLINQ is its ability to decompose a query into smaller tasks that can be executed concurrently across different threads. PLINQ queries automatically partition the data and perform the queries in parallel where possible. This makes it simpler to leverage the hardware for intensive data-processing tasks.

If we were to rewrite the query in the preceding LINQ code snippet so that it uses PLINQ, it would look like this:

```
var numbers = new List<int> { 1, 2, 3, 4, 5, 6 };
var evenNumbers = from num in numbers.AsParallel()
                  where num % 2 == 0
                  select num;
```

This time, the `AsParallel()` method is used to parallelize the query. This method hints that the PLINQ engine utilizes multiple threads, each processing part of the data.

PLINQ uses a query execution engine that automatically partitions the source data and schedules it onto multiple threads managed by the .NET thread pool. The engine applies various optimization strategies, such as the following:

- **Partitioning**: Dividing the source data into smaller chunks that can be processed independently
- **Merging**: Combining results from individual tasks
- **Aggregation**: Efficiently summarizing data from multiple tasks

These strategies are designed to maximize parallel execution without requiring explicit thread management from the developer. PLINQ provides several mechanisms to manage memory effectively during parallel operations:

- **Thread-local storage**: PLINQ utilizes local storage for threads to minimize contention. For instance, each thread may maintain its buffer to aggregate results, which are merged later.
- **Lazy initialization**: Resources are allocated only when necessary, and their initialization is deferred until the executing thread requires it.
- **Garbage collection**: .NET's GC works concurrently with PLINQ operations to clean up unused or unreferenced objects, thus helping manage memory without manual intervention.

To maximize the effectiveness of PLINQ and manage memory efficiently, developers should follow several best practices:

- **Use AsParallel wisely**: Not all queries benefit from parallelization. Performance can degrade if the overhead of thread management exceeds the benefits. It is crucial to profile and test queries in parallel and sequential forms.
- **Handle exceptions carefully**: PLINQ aggregates exceptions thrown by individual threads and throws them as an `AggregateException` error. Developers must handle these exceptions to prevent memory leaks and ensure resource cleanup.
- **Limit the degree of parallelism**: Overloading the system with too many concurrent threads can lead to excessive memory usage and increased context switching. The `WithDegreeOfParallelism` method can be used to control the number of simultaneous operations.

We will review the concepts of **AggregateException** and the **degree of parallelism** later in the chapter.

PLINQ is a powerful tool in .NET, and understanding how PLINQ manages concurrency and memory allows developers to leverage this technology better to improve the performance of their applications.

Now, let's focus on which collection types best suit concurrent operations.

Concurrent collections

Concurrency in .NET is designed to facilitate the development of applications that manage multiple tasks, leverage multi-core processors, and maintain responsiveness. Developers can implement complex concurrent patterns using built-in frameworks and libraries with less effort and more reliability. .NET provides several thread-safe collections under the `System.Collections.Concurrent` namespace, such as `ConcurrentBag`, `ConcurrentDictionary`, `ConcurrentQueue`, and `ConcurrentStack`. These collections offer safe and efficient ways to handle data in multi-threaded environments without requiring manual synchronization using locks.

Using incorrect collections in concurrent code in .NET can lead to several issues, primarily data integrity, performance, and system reliability. Multiple threads manipulating a non-thread-safe collection simultaneously without proper synchronization can corrupt data. For example, adding items to a `List<T>` type from multiple threads can result in some items not being appropriately added or in an unpredictable collection state.

We also risk a race condition occurring when the system's behavior depends on the sequence or timing of uncontrollable events (such as thread execution order). Using non-thread-safe collections without synchronization mechanisms (such as locks and mutexes) can result in race conditions, where overlapping operations produce incorrect or unexpected results.

Without proper memory barriers, often handled automatically by thread-safe collections, there's a risk of memory inconsistencies. Threads might see stale data due to caching and not accessing the most recent memory state, leading to incorrect behavior or results. Debugging concurrent systems can also be significantly more challenging, increasing the risk of leaving subtle bugs that only occur under specific timing or loads.

To mitigate these issues, .NET provides several thread-safe collections designed to handle multiple threads for adds and removes without requiring additional synchronization. Some of the common thread-safe collections are as follows:

- `ConcurrentDictionary<TKey, TValue>`: Ideal for applications where you need to store key-value pairs that are accessed and updated frequently by multiple threads, such as caching scenarios where different parts of an application need to update a shared resource dictionary.

- `BlockingCollection<T>`: Best for producer-consumer scenarios where threads produce data and other threads consume that data. It is useful when you must implement throttling (limit the collection size to prevent out-of-memory scenarios) or ensure that consumer threads block and wait when no items are available.

- `ConcurrentQueue<T>`: Suitable for scenarios where data needs to be processed in the order it was added. It is commonly used in task scheduling, where operations are queued up to be processed by worker threads sequentially.

- `ConcurrentStack<T>`: Ideal for scenarios where you need last-in, first-out access, such as in undo functionality in applications, or for storing temporary data in scenarios where the most recent data needs to be accessed first.

- `ConcurrentBag<T>`: Particularly useful in scenarios where the order of items is not important, and tasks need to grab an object to process. This collection is often used in scenarios involving task parallelism, where each thread can store and retrieve items independently of other threads.

By way of example, let's imagine that we have one list of integer values that needs to serve multiple tasks. The code snippet for this might look like this:

```
using System;
using System.Collections.Generic;
using System.Threading.Tasks;

var numbers = new List<int>();

Task task1 = Task.Run(() => {
    for (int i = 0; i < 1000; i++) {
        numbers.Add(i);
    }
});

Task task2 = Task.Run(() => {
    for (int i = 0; i < 1000; i++) {
        numbers.Add(i);
    }
});

Task.WaitAll(task1, task2);
Console.WriteLine($"Total count: {numbers.Count}");
    // This may not be 2000!
```

Since the `List<T>` type is not thread-safe, the `Add` method can result in a race condition where elements are added incorrectly or lost, leading to an unexpected count, and potentially corrupting the internal state of the list.

The ConcurrentBag<T> type is a thread-safe collection that's designed for scenarios where multiple threads add and remove items. Here's how we can rewrite our code so that it's thread-safe:

```
using System;
using System.Collections.Concurrent;
using System.Threading.Tasks;

var safeNumbers = new ConcurrentBag<int>();

Task safeTask1 = Task.Run(() => {
    for (int i = 0; i < 1000; i++) {
        safeNumbers.Add(i);
    }
});
Task safeTask2 = Task.Run(() => {
    for (int i = 0; i < 1000; i++) {
        safeNumbers.Add(i);
    }
});

Task.WaitAll(safeTask1, safeTask2);
Console.WriteLine($"Total count: {safeNumbers.Count}");
// This will be 2000
```

In this version, using ConcurrentBag<T> ensures that each addition operation is thread-safe. The internal state of the collection is managed automatically to handle concurrent modifications, preventing data corruption and ensuring that the count of items is as expected.

Finally, switching to a thread-safe collection can significantly reduce the risks associated with concurrent data modifications. Thread-safe collections automatically manage synchronization and make writing safe concurrent code easier without the complexities of manual locking and synchronization. This approach improves safety, code readability, and maintainability.

Now that we've explored several scenarios that govern concurrent programming, let's focus on developing multi-thread applications.

Memory usage in multi-threaded applications

Multi-threading is a powerful technique for improving application performance by allowing multiple operations to run concurrently. This capability is handy in modern computing environments where processors have multiple cores, enabling parallel code execution. However, effective multi-threading requires a deep understanding of memory usage, potential pitfalls, and best practices to avoid common errors.

Multi-threaded programming is a subset of concurrent programming that involves using multiple threads within a single process to perform different tasks simultaneously. Unlike concurrent programming, which is more about structure and could involve a single processor, multi-threading explicitly involves multiple execution threads managed by a multitasking pre-emptive scheduler.

It is well documented that .NET manages memory through a managed heap where all .NET objects reside. The .NET GC automatically allocates and releases memory, reducing the developer's burden. However, multi-threaded applications face unique memory management challenges:

- **Increased memory usage**: Each thread typically requires its own stack space, which increases the application's overall memory footprint

- **Synchronization overhead**: Protecting shared data with locks and other synchronization mechanisms can increase complexity and overhead

- **Contention**: High contention on shared resources can lead to thread thrashing and inefficient memory usage

Choosing between single-threaded and multi-threaded approaches depends on various factors. We will explore some key differences next.

Single-threaded versus multi-threaded approaches

Single-threaded development in C# refers to using one thread of execution within an application. This model is straightforward and remains widely used, particularly for applications with simpler logic or those that do not demand high levels of concurrency or parallelism. Some of the benefits of taking a single-threaded approach in development are as follows:

- **Simplicity**: Single-threaded applications are generally easier to program and debug because you don't have to deal with the complexities of synchronization, race conditions, or deadlocks, all of which are common in multi-threaded environments.

- **Consistency and determinism**: Operations in a single-threaded application are performed one at a time in a specific order, which can make behavior more predictable and more accessible to trace. This deterministic nature simplifies state management and debugging.

- **Reduced overhead**: Avoiding threads saves the overhead of context switching and thread management. There's no need for locking mechanisms or other synchronization primitives, which can consume system resources and degrade performance.

- **Suitable for simple applications**: Single-threaded designs might be sufficient for applications with a linear flow or primarily performing I/O-bound operations, particularly when combined with asynchronous programming techniques to keep the application responsive.

Let's review a straightforward example of a simple application that calculates the sum of an array of numbers. We will demonstrate this using a single-thread approach, where the operation is run sequentially:

```
using System;

public class SingleThreadedSum
{
    public static void Main()
    {
        int[] numbers = { 1, 2, 3, 4, 5, 6, 7, 8, 9, 10 };
        long sum = SumArray(numbers);
        Console.WriteLine($"The sum of the array is: {sum}");
    }

    private static long SumArray(int[] numbers)
    {
        long sum = 0;
        foreach (int number in numbers)
        {
            sum += number;
        }
        return sum;
    }
}
```

A single-threaded approach can be practical for simple tools and applications where complex operations are not required. Although this approach is simple enough to get most jobs done, there is a reason multi-threading is used in several applications. Significant limitations can be imposed on an application that requires performance if a single-threaded approach is used. We might encounter issues such as the following:

- **Scalability limitations**: Single-threaded applications can only run on one CPU core, which limits their ability to scale with additional hardware resources. These applications may not fully use modern multi-core processors as computational demands increase.

- **Performance bottlenecks**: Long-running tasks can block the thread in a single-threaded application, leading to delays in processing other tasks. This can affect applications that require high responsiveness or need to handle multiple tasks or requests concurrently.

- **Poor use of modern hardware**: Most modern systems have multiple cores designed to handle several threads simultaneously. Single-threaded applications cannot take full advantage of these hardware capabilities, potentially underutilizing the available computational power.

- **Responsiveness issues**: A single-threaded model can cause the UI to become unresponsive while processing data or performing lengthy operations. Although asynchronous programming can somewhat mitigate this issue, it may not be sufficient for intensive computations that would benefit from parallel execution.

Multi-threaded development, on the other hand, enables applications to perform multiple operations simultaneously by utilizing multiple threads within a single process. This approach can significantly enhance the performance and responsiveness of applications, especially those designed to run on modern multi-core processors. Some immediate benefits that can be realized are as follows:

- **Improved performance**: Multi-threading allows an application to perform multiple operations concurrently and use the CPU's power more efficiently.

- **Increased responsiveness**: Multi-threading can keep the UI responsive during long-running background tasks. Separate threads can facilitate either operation, preventing the main thread from getting blocked.

- **Better resource utilization**: Modern computers come with multi-core processors, and multi-threading enables applications to leverage these cores. This leads to better resource utilization and can significantly enhance performance on multi-core systems.

- **Scalability**: Multi-threaded applications can handle more tasks in less time. This scalability is crucial for server-side applications that handle multiple client requests simultaneously.

Now, if we were to take a multi-threaded approach to the previous example and calculate the sum of an array using multiple threads, the code might end up looking like this:

```
using System;
using System.Collections.Generic;
using System.Threading;

public class MultiThreadedSum
{
    static void Main()
    {
        int[] numbers = { 1, 2, 3, 4, 5, 6, 7, 8, 9, 10 };
        int numberOfThreads = 4; // Be wise when choosing this number.
        long sum = SumArrayMultiThreaded(numbers, numberOfThreads);
        Console.WriteLine($"The sum of the array is: {sum}");
    }

    private static long SumArrayMultiThreaded(int[] numbers,
    int numberOfThreads)
    {
        int lengthPerThread = numbers.Length / numberOfThreads;
        long totalSum = 0;
```

```
        List<Thread> threads = new List<Thread>();
        object lockObject = new object();

        for (int i = 0; i < numberOfThreads; i++)
        {
            int start = i * lengthPerThread;
            int end = (i == numberOfThreads - 1) ?
            numbers.Length : start + lengthPerThread;
            Thread thread = new Thread(() =>
            {
                long partialSum = 0;
                for (int j = start; j < end; j++)
                {
                    partialSum += numbers[j];
                }

                lock (lockObject)
                {
                    totalSum += partialSum;
                }
            });

            threads.Add(thread);
            thread.Start();
        }

        foreach (var thread in threads)
        {
            thread.Join();
            // Ensuring all threads complete before proceeding
        }

        return totalSum;
    }
}
```

This code snippet has several considerations. A lock is used when you're updating the total sum to prevent race conditions where multiple threads could try to update the totalSum variable simultaneously, leading to incorrect results. Threads are also explicitly created. The Join method is used to ensure that the main thread waits for all the worker threads to complete before it proceeds to report the total sum. Finally, while splitting the task among multiple threads can significantly reduce the time required to process the dataset, small datasets don't benefit from this approach when considering the overhead of thread management and the complexity that needs to be introduced.

Ultimately, multi-threading introduces some of the issues discussed previously in concurrent development, where developers must handle synchronization, manage thread life cycle, and ensure thread safety. This complexity increases the likelihood of bugs and can make maintenance more challenging. There can be significant issues because threads often share data and resources. If they are not adequately managed, they can interfere with each other, leading to unpredictable behavior and errors.

Unpredictable behaviors and errors can be considered bugs that need to be fixed, and debugging is the go-to method for repairing these. Consequentially, debugging multi-threaded applications can be challenging because bugs may occur. This means that issues might not surface consistently, making them hard to reproduce and fix. Testing becomes more complex as you must ensure the application behaves correctly under various concurrent execution paths.

We must also consider that each thread requires additional memory for stack space. Overusing threading can lead to resource exhaustion in the target system and reduce performance, which is quite the opposite of the implementation's expectations.

C# provides a robust set of features designed to facilitate multi-threaded programming. These features help developers efficiently manage multiple threads, handle synchronization issues, and ensure thread safety. While we have reviewed some of them in the previous section, we'll review them here:

- **The Thread class**: The `System.Threading.Thread` class is a fundamental feature for creating and managing threads directly. It allows developers to execute code on separate threads.

- **ThreadPool**: The `ThreadPool` class manages a pool of threads that can be reused for multiple tasks, reducing thread creation overhead, especially for short-lived operations.

- **TPL**: TPL simplifies writing multithreaded and asynchronous code, abstracting many complexities of thread management. It uses the `Task` class, which represents an asynchronous operation.

- **Locks and synchronization**: The `lock` keyword and other synchronization primitives, such as `Mutex`, `Semaphore`, and `Monitor`, prevent race conditions and ensure that only one thread can access a resource.

- **async and await**: Freeing up the main thread simplifies handling asynchronous operations and improves application responsiveness. These keywords help write asynchronous code that is as straightforward as synchronous code.

These methods can significantly aid in managing memory in a multi-threaded environment. These techniques help reduce the overhead associated with thread management, ensure safe access to shared resources, and improve overall application stability and performance by managing the complexities of multi-threaded execution patterns.

The `ThreadPool` class, for instance, helps by managing a pool of worker threads that are reused to perform multiple tasks. This reuse avoids the constant overhead of thread creation and destruction, leading to better memory and resource management. This dynamic management helps prevent the system from becoming overwhelmed by too many concurrent threads, which could exhaust memory and processing resources.

Locks help control access to shared resources and prevent multiple threads from modifying the shared state concurrently, which can lead to corrupt data and unpredictable memory usage. This stability is vital for avoiding memory leaks and other issues arising from improperly synchronized access to shared resources.

Whether we use multi-threaded development should be based on the application's specific needs, the expected load, the performance requirements, and the developer's ability to manage complexity.

One standard method of implementing multi-threading in applications is parallel loops. We will review this concept next.

Parallel loops

Traditional loops execute their iterations one at a time in the order they are written. Each iteration must be completed before the next one begins. They are straightforward to implement and understand, making them a good choice for simple tasks or tasks that must be executed in a specific order. The output of traditional loops is predictable and repeatable since iterations are processed in a strict sequence. This means that there is no overhead associated with managing multiple threads, which can be beneficial for tasks that are not CPU-intensive or when the overhead of parallelization would outweigh its benefits.

Here's a common scenario where an operation is performed using a traditional `foreach` loop:

```
List<Data> dataList = GetDataList();
List<Result> results = new List<Result>();

foreach (var data in dataList)
{
    results.Add(ProcessData(data));  // Time-consuming operation
}
```

Since this is single-threaded, one thread will be locked in this operation until completion. Using a multi-threaded approach, we can distribute the work to additional threads and use more resources more efficiently for a faster process.

A parallel loop can execute multiple concurrent iterations, utilizing numerous processor cores simultaneously. Distributing the workload across different threads/cores can significantly reduce the time required to complete tasks, which is especially beneficial for large datasets or CPU-intensive tasks. Parallel loops tend to be more complex due to potential issues such as race conditions, deadlocks, and managing thread safety. They can also be more resource-intensive, requiring more CPU power and increasing contention and overhead associated with thread management.

Parallel loops are a part of TPL, providing a powerful parallel programming model. These loops allow for the concurrent execution of code across multiple threads, making it easier to perform operations that can be divided into independent tasks that run simultaneously. The most used parallel loops in C# are `Parallel.For`, `Parallel.ForEach`, and `Parallel.Invoke`. Each of these constructs is designed to handle different scenarios and types of data.

The `Parallel.For` construct is used for parallelizing operations represented by a `for` or counter-controlled loop, where the iterations are independent. Each loop iteration can run concurrently, which is ideal for CPU-bound tasks, and can be executed in any order.

The `Parallel.ForEach` construct works similarly to `Parallel.For` but is used to iterate over any `IEnumerable<T>` type, making it more versatile for operations on collections that are not inherently index-based. It is handy for processing items in a collection in parallel. Here's an example of using `Parallel.ForEach` to replace the previous single-threaded approach with a `foreach` loop:

```
List<Data> dataList = GetDataList();
ConcurrentBag<Result> results = new ConcurrentBag<Result>();
Parallel.ForEach(dataList, data =>
{
    Result result = ProcessData(data);   // Time-consuming
    results.Add(result);   // No need for external locking
});
```

The `Parallel.ForEach` method distributes the processing load across multiple threads and potentially multiple cores. This can significantly reduce the time required to process all data items compared to a sequential approach. The `ConcurrentBag<T>` type is used in place of the `List<T>` type since it is a thread-safe collection suited for situations where the order of items doesn't matter. It allows multiple threads to add items concurrently. This modification will help maintain high performance while ensuring the code remains thread-safe.

The `Parallel.Invoke` construct is used to execute multiple actions in parallel. It does not iterate over a collection but rather starts several tasks at once. This is useful when you have a set of operations that can be performed simultaneously and independently. An example of this is shown here:

```
using System;
using System.Threading.Tasks;

public class ParallelInvokeExample
```

```
{
    public static void Main()
    {
        Parallel.Invoke(
            () => Console.WriteLine("Task 1"),
            () => Console.WriteLine("Task 2"),
            () => Console.WriteLine("Task 3")
        );

        Console.WriteLine("All tasks completed.");
    }
}
```

The `Parallel.Invoke` construct executes three different tasks concurrently. This approach is useful for kicking off multiple unrelated operations that must be run simultaneously.

Since parallel loops use multiple threads, we must be mindful that we want to avoid thread exhaustion while the application runs. We must also ensure that we limit the number of threads that can be used. Parallel loops support a `ParallelOptions` parameter, which contains a property called `MaxDegreeOfParallelism`. This helps us limit the number of threads accordingly. Reusing the previous example where we iterate the list with the `Parallel.ForEach` loop, we can limit the number of threads to two, like this:

```
using System.Collections.Concurrent;
List<Data> dataList = GetDataList();
ConcurrentBag<Result> results = new ConcurrentBag<Result>();

// Create an instance of ParallelOptions
ParallelOptions options = new ParallelOptions();
options.MaxDegreeOfParallelism = 2;
// Adjust the degree of parallelism as needed

// Execute the Parallel.ForEach loop with the specified options
Parallel.ForEach(dataList, options, data =>
{
    Result result = ProcessData(data);
    results.Add(result);
});
```

While parallel loops can significantly improve performance for operations, they also introduce the following issues:

- **Error handling**: Handling exceptions becomes more complex because multiple threads might throw exceptions simultaneously. You might need to consider using a `try-catch` block within the loop and managing exceptions appropriately.

- **Resource usage**: It can also lead to higher CPU and I/O usage, which might not always be desirable, especially in resource-constrained environments or when high concurrency might impact other applications.

- **Order of execution**: This does not guarantee that operations are performed in the order of the array elements. If order is necessary, stick with a traditional `foreach` loop.

By understanding the capabilities and appropriate use cases for each type of parallel loop, developers can effectively leverage the power of modern multi-core processors to enhance the performance of their applications. A recurring theme has been the challenge of handling exceptions in concurrent and multi-threaded applications. We'll review some best practices next.

Exception handling in multi-threaded applications

Exceptions are anomalies that occur during the execution of a program and disrupt its normal flow. Handling these anomalies, commonly known as exception handling, is critical to developing robust, reliable software. An exception can be classified as any error or unexpected behavior arising while a program runs. They can be caused by various factors and range from general to specific situations when dealing with customized code and external libraries. A simple example of a general problem is attempting to divide by zero.

Exceptions are objects derived from the `System.Exception` class and can fall into one of two general categories:

- **System exceptions**: The CLR automatically throws these when errors occur during program execution (for example, `DivideByZeroException`)

- **Application exceptions**: These are defined by developers to handle errors specific to the application's logic

The base class includes several properties that help identify the cause and location of an error:

- `Message`: A description of the error

- `StackTrace`: Shows where the exception occurred

- `InnerException`: Stores the original exception in cases where exceptions are rethrown with additional contextual data

The `try-catch` block is the simplest method for handling exceptions. The code that might throw an exception is placed inside the `try` block, and the code that handles the exception is written in the `catch` block. A `try` block can have multiple catch blocks to handle different exceptions differently. Always order `catch` blocks from the most specific to the least specific to ensure that all exceptions are handled correctly.

An example is shown here:

```
try
{
    int divisor = 0;
    int result = 10 / divisor;
}
catch (DivideByZeroException ex)
{
    Console.WriteLine($"Error: {ex.Message}");
}
catch (Exception ex)
// Catch-all block for any other type of exception
{
    Console.WriteLine("General error: " + ex.Message);
}
```

Every time an exception is thrown, an exception object is created. In scenarios where exceptions are thrown frequently (such as in a loop or frequently used methods), the memory overhead can be significant since each new object requires allocation, which, if not appropriately managed, can lead to the following:

- **Heap pollution**: A rapid accumulation of poorly handled exception objects on the heap can trigger more frequent garbage collection.

- **Generation promotion**: If exception handling is poorly implemented – for example, exceptions are stored in long-lived data structures – exception objects may be promoted to older generations in the GC and collected less frequently, leading to higher memory usage over time.

- **Unreleased memory**: Objects or resources that were supposed to be released or reset in the normal flow might remain allocated or open when an exception is thrown, leading to memory leaks.

- **Data corruption**: Improper handling of exceptions can lead to partial updates to data structures, which might consume excess memory and leave the application inconsistent.

To mitigate the negative impacts of exceptions on memory management, exceptions should be used for exceptional conditions and not regular control flow. This means we should not use exceptions to handle general logic and only use them in situations affecting the application's programmatic runtime.

We should also consider a finally block in the try-catch statement. This allows us to write additional logic to clean up resources that might linger after the exception is thrown. Previously in this book, we saw an example of using a finally block in the implementation of the IDisposable pattern. Here's an example of us attempting to open and read the contents of a file that doesn't exist. This leads to an exception, which is caught, but more importantly, the resource is disposed of after the operation:

```
using System;
using System.IO;

class Program
{
    static void Main()
    {
        FileStream fileStream = null;
        try
        {
            // Attempt to open a file that doesn't exist
            fileStream = File.Open("example.txt", FileMode.Open);
            // Read data from the file
        }
        catch (FileNotFoundException ex)
        {
            Console.WriteLine("File not found: " + ex.Message);
        }
        catch (Exception ex)
        {
            Console.WriteLine("An exception has occurred:
" + ex.Message);
        }
        finally
        {
            // Ensure that the file stream is closed properly
            if (fileStream != null)
            {
                fileStream.Close();
            }
        }
    }
}
```

The finally block is executed whether or not an exception is thrown. It checks if fileStream is not null, indicating that the file was opened successfully but may or may not have been read, and closes the file stream to release the file handle.

By understanding and addressing the impact of exceptions on memory management, developers can enhance the performance and reliability of their C# applications.

The `AggregateException` exception handles exceptions from multiple threads or tasks in parallel programming and task-based operations. This is an essential concept since different tasks or threads might throw exceptions, and they will not be reported individually. The developer must iterate through the different types of exceptions thrown and take appropriate action. The `AggregateException` exception is typically encountered in the following contexts:

- **TPL**: When using `Task` and `Task<T>` in asynchronous operations, if multiple tasks throw exceptions, `Task.Wait()` or `Task.Result` will throw an `AggregateException` exception that contains all the exceptions from the tasks

- **Parallel programming**: When using parallel loops, if multiple iterations throw exceptions, they are aggregated and thrown as a single `AggregateException` exception

Effectively handling an `AggregateException` exception requires understanding the structure and handling each contained exception appropriately. The most straightforward approach involves catching `AggregateException`, calling `Flatten()` to deal with nested `AggregateException` exceptions, and iterating over `InnerException` exceptions:

```
try
{
    Task.WaitAll(tasks); // Execute several tasks
}
catch (AggregateException ex)
{
    foreach (var innerEx in ex.Flatten().InnerExceptions)
    {
        Console.WriteLine(innerEx.Message);
    }
}
```

You may also consider handling only those exceptions to which you can meaningfully react. C# exception filters can selectively catch exceptions based on specific criteria, reducing the complexity within `catch` blocks. You can also consider logging to capture other unrelated exceptions and their inner exceptions for later debugging.

Handling `AggregateException` exceptions requires a thoughtful approach tailored to the concurrency model of your application. Now that we have explored multi-threading with parallel programming, let's focus on asynchronous programming.

Memory management in asynchronous code

Asynchronous programming is not to be confused with multi-threading or concurrency. This programming method allows developers to perform resource-intensive operations without blocking the main thread of execution, thereby improving applications' responsiveness. The `async` and `await` keywords in C# are primarily used to define asynchronous methods that can perform operations such as file I/O, network requests, or intensive computations in a non-blocking manner.

Concurrent, multi-threaded, and asynchronous programming are concepts often used to improve application performance and responsiveness. Although they may seem similar at first glance, each has distinct characteristics and is suited to solving different problems in software development.

Let's recall that concurrent programming refers to the ability of an application to make progress on more than one task at the same time. It doesn't necessarily mean that tasks are executed simultaneously; instead, it involves structuring the application so that multiple tasks can be worked on within overlapping periods. This is achieved by interleaving the execution of tasks broken into smaller sub-tasks that can be executed out of order without affecting the outcome. Multi-threaded programming is a subset of concurrent programming that specifically involves using multiple threads within a single process to perform different tasks simultaneously.

Asynchronous programming, while related to concurrency, primarily deals with handling tasks without waiting for them to complete before moving on to the next task. This means that a program can start a task and immediately move on to another task without waiting for the first one to complete. This keeps the application responsive by freeing up the main execution thread to do other work while waiting for I/O operations or other long-running tasks to complete.

The `async` and `await` keywords were introduced in .NET 4.5, providing a powerful asynchronous programming model. Unlike traditional threading and task-based approaches, async and await allow developers to write non-blocking code that resembles synchronous code in its structure. This simplifies handling I/O-bound tasks, such as file access, network calls, or database transactions, allowing them to run without blocking the main thread and thus keeping UIs responsive and services agile. Let's review how we can create and use asynchronous methods.

Creating and using asynchronous methods

Since asynchronous programming aims to increase performance and better use resources, converting a synchronous method into an asynchronous one is vital. The first step is to identify I/O operations such as file access, database queries, or network calls, but it can also include intensive computational tasks. Once these operations have been identified, we can modify the code to enable the method to operate without blocking the thread it runs on.

Next, we can look for asynchronous equivalents of these blocking operations. Several I/O-bound APIs in .NET provide asynchronous methods. Some examples are as follows:

- Use `ReadAsync`, `WriteAsync`, `CopyToAsync`, and others instead of their synchronous counterparts for file-based operations.

- Use asynchronous versions of commands and queries, such as `ExecuteReaderAsync` or `ExecuteNonQueryAsync`. If you're using Entity Framework Core, consider `SaveChangesAsync` instead of `SaveChanges`.

- Use HttpClient's `GetAsync`, `PostAsync`, and other methods for performing API operations.

Once you've started using the asynchronous versions of the methods, you need to modify the method so that it can be called asynchronously in other parts of the code base. Change the method's return type to `Task` if it was a void method or `Task<T>` if it returns a value. Here, T represents the original return type. Also, add the `async` keyword to the method signature. This keyword enables the use of `await` within the method.

For instance, the following synchronous method, which opens and reads a file's contents, would be a perfect candidate for conversion:

```
public string ReadFile(string filePath)
{
    using (var reader = new StreamReader(filePath))
    {
        return reader.ReadToEnd();
    }
}
```

We can use the `ReadToEndAsync` method, change the method signature, and introduce the `await` keyword like this:

```
public async Task<string> ReadFileAsync(string filePath)
{
    using (var reader = new StreamReader(filePath))
    {
        return await reader.ReadToEndAsync();
    }
}
```

If no asynchronous alternative is available, we must refactor the existing code. We can use `Task.Run` to run the synchronous operation in a background thread, which keeps the primary thread responsive. However, this is generally recommended for CPU-bound operations rather than I/O-bound operations, as I/O-bound should ideally be naturally asynchronous. A simple example of such a legacy method is shown here:

```
public int CalculateResult()
{
    // Blocking operation
    return ExpensiveCalculation();
}
```

We can convert this method into an asynchronous method like so:

```
public async Task<int> CalculateResultAsync()
{
    // Non-blocking operation
    return await Task.Run(() => ExpensiveCalculation());
}
```

Instead of returning the value, we return the result of the `Task.Run` method, which invokes the functionality on another thread. Optionally, we can rename the method and add `Async` as a suffix, which helps with the code's readability.

Now that we know how to convert methods into asynchronous methods, we can review some practical implementations in an application. Consider the following synchronous operation. It loads several files, which can be a time-consuming operation. As a synchronous operation, we can expect that the main thread will be locked into the operation, and the quality of the user experience will be reduced:

```
public void ProcessFiles(string[] filePaths)
{
    foreach (var path in filePaths)
    {
        try
        {
            LoadFile(path);   // Synchronous file loading method
        }
        catch (Exception ex)
        {
            Console.WriteLine($"File processing threw an exception:
            {ex.Message}");
        }
    }
}
```

```
private void LoadFile(string path)
{
    // Synchronously read the contents of a file
    System.IO.File.ReadAllText(path);
}
```

The most significant disadvantage of this synchronous method is its performance. In the asynchronous version, multiple files could be loaded in parallel, utilizing I/O concurrency and potentially multi-core processors to handle I/O-bound operations faster. The synchronous version processes each file one at a time, which can significantly increase the total processing time, especially with many files.

Asynchronous methods are defined using the async keyword, and the return type is usually Task or Task<T>, where T represents a generic type. When an async method is called, it returns a Task type representing the ongoing operation, allowing the calling thread to continue executing without blocking. Here is an asynchronous version of the same method, which uses a Task-based asynchronous programming approach to operate without blocking the main thread:

```
public async Task ProcessFilesAsync(string[] filePaths)
{
    var tasks = filePaths.Select(path =>
    LoadFileAsync(path)).ToArray();
    try
    {
        await Task.WhenAll(tasks);
    }
    catch (AggregateException ex)
    {
        foreach (var innerEx in ex.Flatten().InnerExceptions)
        {
            Console.WriteLine($"Task threw exception:
            {innerEx.Message}");
        }
    }
}


private async Task LoadFileAsync(string path)
{
    // Asynchronously read the contents of a file
    string content = await System.IO.File.ReadAllTextAsync(path);
}
```

The `async` keyword declares that the method will handle its operations asynchronously, which is essential for the `await` keyword to work. Similarly, the `await` keyword indicates that we want to return to the code and execute everything after it once the data is loaded. Here, `Task.WhenAll` is used to await multiple tasks, which is especially useful when you have multiple asynchronous operations that can run concurrently and want to wait for all of them to complete.

Then, `ReadAllTextAsync` is used to read files asynchronously. It returns a `Task<string>` type, which completes after the file reading operation. The `await` keyword is used to asynchronously wait for the operation to complete without blocking the executing thread.

Finally, the `try-catch` block is used to catch exceptions from any tasks. As discussed previously, `AggregateException` will be unwrapped, so the original exceptions are directly caught and processed.

Now that we know how to author asynchronous methods, let's review some best practices and techniques to prevent memory leaks as we advance.

Preventing memory leaks and best practices

When you write an asynchronous method using `async` and `await`, the C# compiler generates a *state machine* behind the scenes. This state machine handles the control flow of the method, allowing it to pause and resume execution around `await` statements. The state machine captures the local variables, parameters, and the current state of the method. While this transformation simplifies writing asynchronous code, it can introduce subtle memory leaks if not managed carefully. Understanding how the state machine works and how it can lead to memory leaks is essential for writing efficient and reliable asynchronous code.

One primary way the state machine can cause memory leaks is by capturing references to local variables, parameters, and the `this` reference. If these captured references include large objects or objects with long lifetimes, they can prevent the GC from reclaiming memory.

Take, for instance, the following example, where a data object might get caught in a long-running async operation:

```
public class DataProcessor
{
    private byte[] _data;

    public DataProcessor(byte[] data)
    {
        _data = data;
    }

    public async Task ProcessDataAsync()
    {
        await Task.Delay(1000); // Simulate asynchronous work
```

```
        Console.WriteLine(_data.Length);
    }
}
```

Here, we can see that the asynchronous `ProcessDataAsync` method references `_data` directly, which causes it to also capture the `this` reference. We simulate a long process, but if the method is long-running or if instances of `DataProcessor` are retained longer than necessary, this can lead to memory leaks. A suitable alternative solution to this issue would be to pass the data as a parameter to the `async` method. This way, there is no direct binding to the object's field, and the variable can be cleaned up at the end of the long-running process, causing a memory leak:

```
public class DataProcessor
{
    private byte[] _data;

    public DataProcessor(byte[] data)
    {
        _data = data;
    }

    public static async Task ProcessDataEfficientlyAsync(byte[] data)
    {
        await Task.Delay(1000).ConfigureAwait(false);
        // Simulate asynchronous work
        Console.WriteLine(data.Length);
    }
}
```

Notice that in addition to using a parameter variable, we also introduced the `ConfigureAwait()` method. This instructs the runtime not to capture the current *synchronization context*, thus not forcing the continuation to run on the original context. This will help avoid unnecessarily capturing the synchronization context, which is instrumental in library code where you do not need to marshal the continuation back to the original context.

The primary role of `SynchronizationContext` is to provide a generic mechanism for synchronizing asynchronous callbacks and message posting to a specific thread. In a typical UI application, the default `SynchronizationContext` ensures that continuations run on the UI thread.

Applications built with Windows Forms or WPF that feature a UI, for example, use `SynchronizationContext` to ensure that UI updates are executed on the main UI thread. This is necessary because most UI components are not thread-safe and must be updated on the thread they were created on.

One of the first memory management challenges is context capturing. When an asynchronous method awaits a task, the current `SynchronizationContext` uses it to resume the process after the await. This benefits UI applications where operations need to return to the UI thread. However, it can inadvertently lead to holding references to UI controls or other objects longer than necessary, increasing memory usage and potentially leading to memory leaks.

When an asynchronous operation is started from a thread associated with a synchronization context (such as the UI thread in a Windows Forms application), the `SynchronizationContext.Current` property captures the current context, which is then used to marshal the callbacks or continuations to the originating thread once the asynchronous operation is complete. A simplified sequence of events is as follows:

1. An asynchronous operation is initiated.

2. The `SynchronizationContext.Current` property captures the context of the current thread, if available.

3. The asynchronous operation runs on a different thread.

4. The continuation/callback is posted to the captured `SynchronizationContext` when the task is completed.

5. The `SynchronizationContext` class executes the continuation on the original thread.

While `SynchronizationContext` is beneficial for orchestrating asynchronous tasks, it can also introduce complications:

- **Performance overheads**: Capturing and marshaling continuations to the original context can lead to performance issues if the UI thread, for example, is busy, making the application less responsive.

- **Deadlocks**: One of the most common issues arises when the UI thread waits on a task using `Task.Wait()` or `Task.Result`, and that task awaits another operation, which attempts to post its continuation back to the UI thread. If the UI thread is blocked waiting for the task to complete, it cannot process the continuation, leading to a deadlock.

- **Unnecessary context switching**: In a non-UI or server-side application, there might be no need to marshal callbacks to the original thread. This can lead to unnecessary overhead associated with default context switching and degrade performance.

To address this, developers can use `ConfigureAwait(false)` when awaiting tasks.

In the following code, we have a method that fetches data from an API asynchronously. It is part of a library used by UI and non-UI applications such as WPF or Windows Forms. It does not need to be marshaled back to the original context, so using `ConfigureAwait(false)` will reduce the overhead of returning to the synchronization context:

```
public async Task<string> FetchDataFromApiAsync(string url)
{
    using (HttpClient client = new HttpClient())
    {
        // Using ConfigureAwait(false) to avoid capturing
        // the synchronization context
        string result = await client.GetStringAsync(url).
        ConfigureAwait(false);
        return result;
    }
}
```

Because `ConfigureAwait(false)` is used, the continuation does not need to be marshaled back to the original context. This reduces the memory allocation and orchestration overhead and eliminates the risk of a deadlock with the calling thread waiting for the task to complete while also needing to execute its continuation.

We can also take measures to avoid blocking calls in UI applications by avoiding the use of `Task.Wait()` or `Task.Result` on asynchronous operations. These two methods force the task to run synchronously, leading to the blocks we attempt to avoid. When you call `Wait()` or `Result` on a task in a UI thread, and the task execution requires executing something back on the UI thread, this continuation will never execute because the UI thread is blocked. The task will wait for the UI thread to be accessible to process the continuation while the UI thread waits for the task to complete, thus creating a **deadlock**.

Blocking calls using `Wait()` and `Result` wastes resources by idling the current thread while other work could have been processed. Using the thread for other tasks is always more efficient than blocking it and waiting for a result. This is especially true in high-performance or highly scalable applications, where managing resources efficiently is critical.

Instead, we should use `await` whenever possible. The `await` method allows the current method to return immediately, freeing the thread to do other work. When the awaited task is completed, the method's execution resumes using a continuation. As seen in the preceding code example, the call to the `HttpClient` method, `GetStringAsync()`, is being awaited.

Splitting large methods into small parts is another relatively simple approach that can make async programming more efficient. Since a method should never be too large or responsible for multiple tasks, this is a double-win regarding code organization and overall efficiency. Splitting large asynchronous methods into smaller methods can reduce the scope of the captured state and make it easier to manage memory:

```
public async Task ProcessDataInPartsAsync(byte[] data)
{
    await PartOneAsync(data).ConfigureAwait(false);
    await PartTwoAsync(data).ConfigureAwait(false);
}

private static async Task PartOneAsync(byte[] data)
{
    await Task.Delay(1000).ConfigureAwait(false); // Simulate part one
}

private static async Task PartTwoAsync(byte[] data)
{
    await Task.Delay(1000).ConfigureAwait(false); // Simulate part two
}
```

Another crucial construct, introduced in .NET Core 2.0, is the ValueTask type. It is a more efficient alternative to Task for certain asynchronous programming scenarios. It can be awaited like Task but offers more flexibility and potentially lower overhead. The ValueTask and ValueTask<T> types are defined in the System.Threading.Tasks namespace.

In practice, tasks allocate an object on the heap every time an asynchronous method is called, regardless of whether it completes synchronously or asynchronously. In contrast, ValueTask can avoid heap allocation by representing the result directly if the method completes synchronously. It can wrap either a Task type or a result, reducing the need for allocations. Here's an example:

```
public async Task<int> GetNumberAsync()
{
    return await Task.FromResult(42); // Allocates a Task<int>
}

public ValueTask<int> GetNumberValueTaskAsync()
{
    return new ValueTask<int>(42);
    // No allocation if completed synchronously
}
```

A ValueTask type is best used in performance-critical paths where avoiding allocations is beneficial. Also, if there is a method that frequently completes synchronously and you want to avoid the overhead of allocating a Task type. While it is an excellent alternative to a task, a task object can be awaited multiple times and always refers to the same asynchronous operation. The ValueTask type, on the other hand, should only be awaited once. Awaiting ValueTask multiple times or using it after it has been awaited can lead to undefined behavior:

```
public async Task<int> GetNumberMultipleAwaitTaskAsync()
{
    var task = Task.FromResult(42);
    int firstAwait = await task;
    int secondAwait = await task; // Safe to await multiple times
    return firstAwait + secondAwait;
}

public async ValueTask<int> GetNumberMultipleAwaitValueTaskAsync()
{
    var valueTask = new ValueTask<int>(42);
    int firstAwait = await valueTask;
    int secondAwait = await valueTask;
    // Undefined behavior, should avoid this
    return firstAwait + secondAwait;
}
```

Generally speaking, avoid ValueTask if the method needs to be awaited multiple times or is used in complex patterns. Another consideration is that when APIs are exposed to a broader audience, Task might be more appropriate due to its simplicity and safety in typical use cases.

As with all programming constructs that we have explored until now, there are pros and cons, as well as guardrails that we should be aware of in implementing our solutions. Thankfully, C# makes asynchronous development relatively straightforward, and if we follow specific patterns and principles, we can implement responsive applications with little effort. Now, let's wrap up this chapter.

Summary

This chapter covered memory management in concurrent, multi-threaded, and asynchronous programming paradigms. Each section provided practical insights and a comprehensive understanding tailored to modern software development's specific demands and complexities in .NET.

We reviewed how .NET handles memory in environments where multiple threads operate in parallel and how the GC seeks to manage memory efficiently and concurrently, focusing on minimizing pause times and maximizing throughput. We discussed concurrent programming challenges such as race conditions, deadlocks, and starvation, as well as solutions such as locking mechanisms, using immutable objects, and employing concurrent collections provided by .NET to mitigate synchronization overhead and improve memory management.

We also examined how multi-threading enables applications to perform multiple operations simultaneously, enhancing performance but introducing complexities in memory management. We then examined strategies for overcoming challenges such as minimizing locked time and choosing the right locking constructs to prevent deadlocks and reduce memory waste.

Finally, we looked at how C# handles asynchronous programming and how memory is managed more efficiently by preventing blocking calls, leading to enhanced application scalability and responsiveness. We also looked at practical examples of converting synchronous methods into asynchronous ones, illustrating how to handle I/O-bound operations without blocking threads, thereby optimizing memory usage.

By integrating these advanced techniques into real-world scenarios, developers can tackle complex memory management issues more confidently and efficiently. This chapter aimed to equip developers with the knowledge to apply advanced memory management techniques effectively in their .NET applications, ensuring that applications are efficient, robust, and scalable.

In the next chapter, we will discuss Memory Profiling and Optimization.

6

Memory Profiling and Optimization

Memory profiling is the process of analyzing a program's memory consumption while it is running. This analysis helps identify where and how memory is used, which parts of the code are memory-intensive, and what triggers memory allocation. Performance optimization in software development often sparks strong opinions among programmers. On one side, an excessive focus on optimizing performance can lead developers down time-consuming rabbit holes, potentially derailing projects and causing delays. Conversely, even minor issues can transform into significant bottlenecks or even more severe problems when multiplied across large systems.

In .NET, this involves analyzing applications' memory usage while running. This process is crucial for identifying how memory is allocated across different parts of a .NET application and understanding the behaviors that lead to excessive memory use or leaks. Effective memory profiling helps developers pinpoint inefficiencies, optimize memory usage, and ensure application stability and scalability.

This chapter will cover various tools and techniques for effective memory profiling. We will explore platform-specific and cross-platform tools that provide deep insights into memory allocation and usage patterns. Each tool will be discussed regarding its features, use cases, and how it fits into the debugging and optimization workflow.

We will review the following topics:

- Memory profiling concepts
- Profiling memory usage and allocation
- Writing unit tests for memory leaks
- Production and deployment considerations

Additionally, this chapter will address advanced topics such as the impact of memory optimization on cloud-based and distributed .NET applications, how to automate memory tests, and the integration of memory profiling into the continuous integration and deployment pipelines.

Let us begin by reviewing memory profiling concepts and techniques.

Technical requirements

- Visual Studio 2022 (`https://visualstudio.microsoft.com/vs/community/`)

- Visual Studio Code (`https://code.visualstudio.com/`)

- .NET 8 SDK (`https://dotnet.microsoft.com/en-us/download/visual-studio-sdks`)

- Git tools (`https://www.git-scm.com/`)

Memory profiling concepts

Memory profiling involves analyzing an application's usage during runtime to identify how memory is allocated, used, and released. The goal is to detect inefficient memory use, such as leaks, excessive memory allocation, and unnecessary memory consumption that can lead to application slowdowns or crashes. It is a critical practice in software development aimed at understanding and optimizing how applications allocate and manage memory. Given the complexities of modern software systems and the variety of hardware environments in which they operate, memory profiling helps ensure applications perform efficiently and reliably.

As we have been exploring since *Chapter 1* of this book, common memory usage issues that we might encounter during application runtime include the following:

- **Memory leaks**: These occur when a program fails to release memory that is no longer needed, decreasing the available memory and potentially causing system crashes over time

- **Allocation overhead**: Involves the extra memory and processing costs associated with allocating and deallocating memory

- **Fragmentation**: Occurs when memory is allocated and deallocated so that small gaps between allocated memory blocks are left unusable

Effective memory profiling should be an integral part of the development process, not an afterthought. This section provides strategies for integrating memory profiling into different stages of software development, from development to testing and maintenance.

.NET applications often run in resource-intensive environments such as web servers and enterprise systems. Optimizing performance helps ensure these applications use system resources, such as memory and processor time, more efficiently. This can reduce operating costs and improve the scalability of the applications. Inefficient memory usage or CPU consumption can lead to application instability and crashes. In .NET, understanding memory management and optimizing garbage collection is crucial to maintaining system stability, especially in long-running applications such as services or cloud-based applications.

As .NET applications scale, small performance inefficiencies can become magnified, impacting the overall system performance. Efficient coding and optimization practices are essential to ensure that the application continues to perform well as load increases without requiring proportional increases in hardware resources. Responsiveness and speed are directly tied to user satisfaction for client-facing .NET applications, such as those built with ASP.NET or .NET MAUI for mobile and desktop. Performance optimization can significantly enhance the user experience by reducing load times and improving the application's responsiveness.

With the increasing adoption of cloud platforms like Azure, performance optimization can translate directly into cost savings. In cloud environments, resources are often billed based on usage. Optimizing the performance of .NET applications can reduce the required compute resources, reducing costs.

These compelling situations encourage developers to take every precaution possible and ensure we profile our applications during development to catch all potential issues. We must understand profiling and profiling tools and how they can be employed. We will review these next.

Understanding and using profiling and tools

Profiling skills are highly valued in the software development industry. Developers proficient in these tools often find adapting to different projects and roles easier, enhancing their career opportunities. When issues arise, proficiency with profiling tools can significantly reduce debugging time. Instead of manually sifting through code or making educated guesses, developers can use these tools to quickly gather empirical data that points directly to the source of the problem. In a market where performance can be a significant differentiator, being adept at using profiling tools can give developers and their companies an edge. Faster, smoother, and more responsive applications increase customer satisfaction and retention.

Profiling is not just for troubleshooting; it's also a proactive measure in the software development lifecycle. Regular use of profiling tools can help developers understand the long-term impacts of their code modifications, guiding them in writing more efficient, cleaner code.

Being comfortable with profiling tools is crucial for developers for several reasons. These tools are pivotal in ensuring software applications' efficiency, performance, and reliability. Profiling tools help pinpoint specific areas where the application may be underperforming. This includes identifying slow or inefficient code paths, excessive CPU usage, memory leaks, or unnecessary database queries. Knowing how to use these tools effectively allows developers to focus their optimization efforts where they will have the most impact.

Memory profiling tools monitor the allocation and deallocation of memory in real-time or through static analysis. These tools can provide visualizations of memory usage over time and detailed reports that break down memory allocation by function, module, or data structure. Profilers often offer the following capabilities:

- **Real-time monitoring:** Tracking memory usage as the application runs, which is crucial for understanding the application's behavior under typical usage conditions

- **Historical analysis:** Analyzing past performance to identify trends and intermittent issues

- **Leak detection:** Identifying points where memory is allocated but not adequately released

- **Allocation profiling:** Showing where and how memory allocations occur within the codebase

For .NET developers, memory profiling is essential due to the managed nature of the environment. .NET's **Common Language Runtime (CLR)** manages memory allocation and deallocation automatically through garbage collection. While this reduces the burden on developers, it also introduces challenges:

- **Understanding garbage collection**: Profiling helps developers understand and optimize the behavior of the garbage collector. For instance, frequent large object heap allocations can lead to performance degradation, which can be identified through profiling.

- **Optimizing memory usage:** .NET developers can use profiling to optimize data structures, manage object lifetimes, and minimize memory footprint, which is especially crucial in performance-critical applications such as games or intensive computational applications.

- **Memory leak detection:** Although managed environments such as .NET reduce the risk of memory leaks, leaks can still occur through event handlers or improper disposal patterns. Memory profiling helps identify these leaks.

A suitable set of profiling tools is crucial to analyze and optimize applications effectively. Here are some of the best profiling tools available, each with its specific strengths and ideal use cases:

- **Visual Studio Diagnostic Tools**: We are not strangers to the tooling available in Visual Studio. These tools provide a seamless experience for profiling .NET applications. They offer a range of features, including CPU, memory, and network usage diagnostics. The integration with the development environment means it's easy to switch between coding and profiling without additional setups. They are a convenient and accessible tool for routine performance checks during development.

- **Redgate ANTS Performance Profiler**: ANTS Performance Profiler is known for its user-friendly interface and powerful profiling capabilities, particularly in identifying slow-running lines of code and database queries. It also integrates with SQL Server, making it an excellent choice for developers working extensively with databases.

- **JetBrains dotMemory**: This tool specializes in memory profiling and offers detailed analysis of memory usage, leaks, and garbage collection. It provides clear insights into how memory is allocated and helps identify objects using too much memory.

- **PerfMon**: Performance Monitor is a versatile system monitoring tool in Microsoft Windows. It provides a wide array of performance counters and metrics that allow users to monitor the health and performance of their computers or servers. The tool can be handy for system administrators, developers, and IT professionals to troubleshoot issues and ensure efficient system operations.

- **Process Explorer**: This powerful utility tool provides detailed information about which files and processes are running on a Windows system. It is often used as an advanced replacement for the Task Manager, offering a more thorough look at the operating system's workings.

Each tool has merits, but each has environmental and cost implications. **Redgate's ANTS Performance Profiler** is currently a premium Windows-only application; and **JetBrains dotMemory,** while being a cross-platform alternative, is also a premium tool that offers a limited free trial. The choice of tool ultimately depends on the project's specific needs, the preferences of the development team, and the issues that need addressing. For instance, if the primary concern is database interaction and its impact on application performance, **Redgate ANTS Performance Profiler** would be a strong choice. Conversely, for general memory management and garbage collection issues, **JetBrains dotMemory** would be ideal.

PerfMon and **Process Explorer** are Windows-based tools that provide insight into the operating system's inner workings. PerfMon is a fundamental tool in the Windows operating environment due to its robustness and flexibility in handling various types of performance data. It can be used for ongoing system maintenance or specific troubleshooting tasks and offers comprehensive capabilities to manage and ensure system performance.

Process Explorer is highly valued by IT professionals, system administrators, and advanced users for its depth of information and versatility in managing Windows processes. It serves as an indispensable tool for deep system analysis and management, providing insights that go far beyond the capabilities of many built-in Windows utilities.

Visual Studio's built-in tooling is the best and most natural choice for memory profiling and monitoring during development. The downside here is that it is only available for Windows machines. Considering that .NET is a cross-platform development framework, it will not be suitable or available for Mac and Linux developers. It is also not a completely free IDE since a license is required to develop commercial and professional solutions.

In *Chapter 4*, we reviewed how memory leaks can be detected using Visual Studio's built-in profiler and looked at a free and cross-platform approach using the dotnet-counters tool provided by the .NET SDK. With this tool, we can visualize memory usage and retrieve object references that might be at the root of excessive memory usage during the application's runtime. Since we have already explored how to use this, we can skip the details here.

Now that we have reviewed some memory profiling concepts and tools and their pros and cons, let us proceed to profiling methods.

Profiling memory usage and allocation

Profiling memory effectively requires several sophisticated techniques, each of which can provide insights into how an application uses memory. Each method serves different purposes and offers unique benefits. Assessing how memory is allocated is vital for understanding a program's memory footprint and behavior. Generally, we need to evaluate the following:

- **Allocation patterns**: Understanding whether memory allocation is static, dynamic, or stack-based helps identify how memory management should be approached

- **Allocation hotspots**: Identifying parts of the code where a high volume of memory allocations occurs can help optimize memory usage and improve application performance

- **Object lifetime management**: Properly managing objects' lifecycles ensures that memory is freed up when it is no longer needed, avoiding memory leaks

Implementing memory profiling effectively involves more than just selecting the right tools. The following best practices are crucial for obtaining accurate results and making informed optimization decisions. Profiling should be practiced as early as possible in the development process to catch potential issues before they become ingrained in the codebase. This proactive approach can save significant time and resources down the line. We must also choose tools that best fit the specific needs of the project and development environment. Consider factors such as the type of application, the deployment environment, and specific performance goals when selecting profiling tools.

Automating the collection and analysis of profiling data can help consistently monitor memory usage across different application lifecycle stages, including testing and production. Once the data is collected, focus on insights that can lead to actionable improvements and prioritize issues based on their impact on application performance and user experience.

Often, we work in teams, and it is good when each team member has some knowledge or experience in making their types of assessments. Education is paramount, and we must ensure that all team members know the principles of memory management and are proficient in using profiling tools. This shared understanding can help prevent memory issues and improve the overall quality of the code. It will also help document the findings from memory profiling sessions and the actions taken to address them, which can help track performance improvements over time. This documentation can also be invaluable for new team members and future references.

Now, let's review techniques for identifying allocation patterns and logging/documenting them with custom code.

Identifying allocation patterns with custom code

Identifying allocation patterns involves understanding how and where memory is allocated for objects and data structures. This can help developers optimize memory usage, improve performance, and reduce issues related to memory leaks and excessive garbage collection.

On the topic of garbage collection, we know by now that this is a marquee feature in .NET development, where the GC automatically frees up memory for us after assessing which objects are still in use and which ones are out of scope. We also know that we can write code to negate the effects of the GC, which is why we end up with memory leaks. Because it becomes difficult to readily and easily detect where the faulty code is being executed, we can add snippets of code to different execution points in our code to try and assess whether garbage collection has occurred and how many times. The following code snippet shows how the GC can be monitored during application runtime:

```
public void MonitorGarbageCollections()
{
    Console.WriteLine($"GC Gen 0 has been collected
    {GC.CollectionCount(0)} times");
    Console.WriteLine($"GC Gen 1 has been collected
    {GC.CollectionCount(1)} times");
    Console.WriteLine($"GC Gen 2 has been collected
    {GC.CollectionCount(2)} times");
}
```

Calling this method at various points in your application can show how frequently different generations of objects are collected, indicating their longevity and allocation rate. Understanding when and why garbage collections occur can provide insights into memory allocation patterns and help to differentiate between short-lived and long-lived objects.

We can also use the GC Notifications API to interact with the garbage collector and memory manager and implement custom memory profiling implementations. We can use `GC.RegisterForFullGCNotification` to get notifications about upcoming garbage collections. This allows an application to receive notifications about impending and completed garbage collections, specifically for full collection cycles. This capability is handy for applications that need to manage memory or resources carefully and can perform optimizations or cleanups before and after a complete garbage collection.

An example of this is as follows:

```
GC.RegisterForFullGCNotification(10, 10);
GCNotificationStatus status = GC.WaitForFullGCApproach();
if (status == GCNotificationStatus.Succeeded)
{
    Console.WriteLine("GC is about to happen. Preparing...");
    // Perform necessary pre-GC operations here
```

```
    }

    status = GC.WaitForFullGCComplete();
    if (status == GCNotificationStatus.Succeeded)
    {
        Console.WriteLine("GC has completed.");
        // Perform necessary post-GC operations here
    }
```

In the preceding code snippet, we start by registering for notifications and providing the `maxGenerationThreshold` and `largeObjectHeapThreshold` parameters. The `maxGenerationThreshold` is the threshold for getting notifications before a complete garbage collection. It determines how aggressively the garbage collector tries to notify the application before starting a full GC. `largeObjectHeapThreshold` is the threshold for post-full garbage collection notifications. The possible values for both parameters range from **1** to **99**, representing the time left before a garbage collection is expected to occur. The lower the value, the more time to respond, and the more frequent checks are needed. Next, we wait for the notification before the GC, and when this is evaluated as `true`, we can execute some code as required. Similarly, we can wait for the completion of the garbage collection and execute code accordingly.

Another code-based method for identifying application activity is logging and instrumentation. You can instrument your code to log memory metrics at critical points, which can help you track down unexpected allocations. This method involves adding custom code to monitor and record information about memory operations, allowing developers to understand memory usage patterns, detect leaks, and optimize performance:

- **Instrumentation**: Modifying the application to include code that measures and records performance and memory usage statistics at crucial stages in the application lifecycle

- **Logging**: This refers to writing the collected data to a log file, a console, or a more sophisticated monitoring system that can be reviewed to understand how the application manages memory over time

.NET provides various APIs to access memory-related information, which can be logged periodically or triggered by specific events or thresholds. The following methods show how the `GC.GetTotalMemory` method logs the total memory the application uses and `GC.CollectionCount` provides the number of times garbage collection has occurred for each generation:

```
public class MemoryProfiler
{
    public static void LogMemoryUsage()
    {
        long totalMemory = GC.GetTotalMemory(false);
        Debug.WriteLine($"Total memory used: {totalMemory} bytes");
```

```
    }

    public static void LogDetailedMemoryUsage()
    {
        for (int i = 0; i <= GC.MaxGeneration; i++)
        {
            long size = GC.CollectionCount(i);
            Debug.WriteLine($"Generation {i} collections: {size}");
        }
    }
}
```

This method logs memory usage before and after a potentially heavy operation, which can help identify unexpected increases in memory usage. To simulate its usage, we can create a sample method to generate a large list and review the usage before and after this heavy operation, as in the following example:

```
public void ExampleMethod()
{
    MemoryProfiler.LogMemoryUsage();
    // Perform memory-intensive operations here
    var largeList = new List<int>();
    for (int i = 0; i < 1000000; i++)
    {
        largeList.Add(i);
    }
    MemoryProfiler.LogMemoryUsage();
    MemoryProfiler.LogDetailedMemoryUsage();
}
```

Here, we use the GC.GetTotalMemory method to log the total memory used by the application, while GC.CollectionCount provides the number of times garbage collection has occurred for each generation. We use the Debug.WriteLine method to emit the diagnostic messages to the output window in Visual Studio and other debugging tools that listen to the debug output stream. It is part of the System.Diagnostics namespace and is a simple and effective tool for developers who need a quick and easy way to inspect what's happening in their code during development, with the assurance that this debug code won't affect the performance or behavior of their application once it's in production.

Now let us jump out of writing custom logs and explore the technique of adding a **Make Object ID** using Visual Studio.

Make Object ID to find memory leaks

During debugging, it is common practice to track a variable to observe its usage and lifetime during the application's execution paths. In Visual Studio, we usually use debugger windows such as the **Watch** window. However, when a variable goes out of scope in the **Watch** window, you may notice it becomes grayed out. In some scenarios, the value of a variable may change even when the variable is out of scope, and the new value(s) cannot be tracked via the debugger **Watch** window. If you must continue watching it closely, you can track the variable by creating an *Object ID* in the **Watch** window. This is where we use the **Make Object ID** feature.

Make Object ID is a valuable feature in Visual Studio that helps track and debug objects in memory, particularly for identifying memory leaks in .NET applications. This identifier remains consistent across different breakpoints and even as you step through the application, allowing developers to track an object's state changes over time and across various parts of the application, regardless of where the object is in the call stack. It's advantageous when objects are passed around through multiple methods or threads.

One of the primary uses of this feature is in detecting memory leaks. By marking an object with an ID, you can quickly check if it remains in memory longer than it should, indicating a potential leak. This is particularly crucial in long-running applications where memory leaks can lead to significant performance degradation or even application crashes over time. When paired with Visual Studio's diagnostic tools, **Make Object ID** can provide a more comprehensive analysis. For instance, after marking an object, you can take memory snapshots at various phases of execution and compare them to see how the object's memory allocation changes, which can help in optimizing memory usage.

Let us consider a typical scenario where the Make Object ID feature can help debug a memory leak involving event handlers in a .NET application. Memory leaks often occur when event handlers are not adequately detached, preventing the garbage collector from reclaiming the memory allocated for objects. In the following code, we define a `EventPublisher` class that raises an event and multiple subscribers (listeners) that attach handlers to this event. Suppose there's a bug causing one of the subscribers not to detach its event handler, leading to a memory leak:

```
public class EventPublisher
{
    public event EventHandler MyEvent;

    public void TriggerEvent()
    {
        MyEvent?.Invoke(this, EventArgs.Empty);
    }
}


public class EventSubscriber
{
```

```
    public void Subscribe(EventPublisher publisher)
    {
        publisher.MyEvent += HandleEvent;
    }

    public void Unsubscribe(EventPublisher publisher)
    {
        publisher.MyEvent -= HandleEvent;
    }

    private void HandleEvent(object sender, EventArgs e)
    {
        Console.WriteLine("Event handled.");
    }
}
```

The `Program.cs` class file contains the following code:

```
var publisher = new EventPublisher();
var subscriber = new EventSubscriber();

// Subscriber attaches to the event
subscriber.Subscribe(publisher);
for (int i = 0; i < 15; i++)
{
    // Simulate event triggering
    publisher.TriggerEvent();

    // Optionally unsubscribe
    // Uncomment the following line to test memory management
    // with unsubscribing
    // subscriber.Unsubscribe(publisher);
} // Keep the console window open
Console.WriteLine("Press any key to exit...");
Console.ReadKey();
```

Now that we have the code, we can place a breakpoint on the line of code that calls the `subscriber.Subscribe(publisher)` method call. Once the code hits the breakpoint, open the **locals** window, right-click on `subscriber` in the list of objects, and select **Make Object ID**. This assigns a unique ID to the subscriber object, let's say {$1}. Review *Figure 6.1* for further insight.

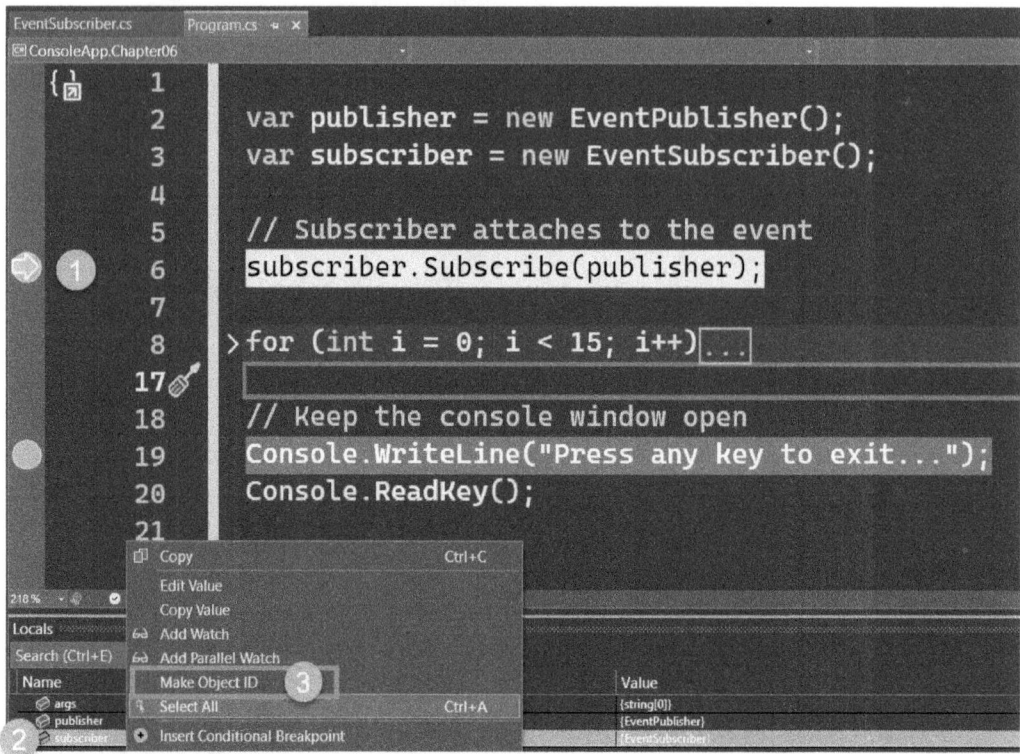

Figure 6.1 – How to add an Object ID to an object during runtime

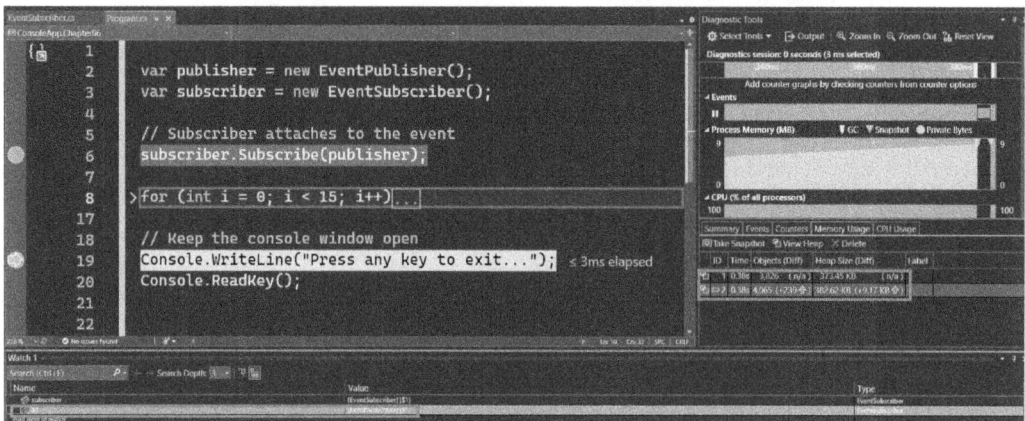

Figure 6.2 – The object id still points to a valid object that should have been collected

The Object ID allows us to directly trace this object's lifecycle and gather concrete proof that it's not being collected as garbage due to lingering event handler references.

Beyond debugging and tracing capabilities in Visual Studio and writing code, we can set up Event Tracing for Windows to gather performance data and diagnostics. We will explore this next.

Event Tracing for Windows

Event Tracing for Windows (ETW) is a high-performance, low-overhead event logging system built into Windows. It is commonly used for performance monitoring, debugging, and tracing application execution. When creating a .NET Core application, you can use ETW to log and trace events. Here's how to set it up and use it, with detailed explanations and code examples.

To use ETW, you need to install the `Microsoft.Diagnostics.Tracing.EventSource` package. This package provides the necessary APIs to create and manage ETW events:

```
dotnet add package Microsoft.Diagnostics.Tracing.EventSource
```

Now that the package is added, we must define the event source. We can use the `EventSource` class as a base class for our custom class and define the events:

```
using System.Diagnostics.Tracing;
[EventSource(Name = "SampleEventSource")]
class SampleEventSource : EventSource
{
    public static SampleEventSource Log {get;}
    = new SampleEventSource();

    [Event(1, Keywords = Keywords.Startup)]
    public void AppStarted(string message)
    => WriteEvent(1, message);
    [Event(2, Keywords = Keywords.Requests)]
    public void RequestStart(int requestId)
    => WriteEvent(2, requestId);
    [Event(3, Keywords = Keywords.Requests)]
    public void RequestStop(int requestId)
    => WriteEvent(3, requestId);
    [Event(4, Keywords = Keywords.Startup,
    Level = EventLevel.Verbose)]
    public void DebugMessage(string message)
    => WriteEvent(4, message);
}


public class Keywords
{
    public const EventKeywords Startup = (EventKeywords)0x0001;
    public const EventKeywords Requests = (EventKeywords)0x0002;
}
```

We define different methods to log a message with a specific log level. A **level** is a number or **LogLevel** string that helps categorize and filter log messages during analysis. The preset levels and leg level strings are as follows:

- 0 = Trace
- 1 = Debug
- 2 = Information
- 3 = Warning
- 4 = Error
- 5 = Critical

Doing this gives us a reusable class and methods to maintain a standard for logging messages in our application. Most event collection and analysis tools use these options to decide which events should be included in a trace:

- **Provider names**: A list of one or more `EventSource` names. Only events defined on `EventSources` in this list are eligible to be included. To collect events from the `SampleEventSource` class above, you must include the `EventSource` name `SampleEventSource` in the list of provider names.

- **Event verbosity level**: Each provider can define a verbosity level, and events with higher verbosity levels will be excluded from the trace. For example, configuring the application to collect **Information** verbosity-level events will exclude **DebugMessage** events since this is higher.

- **Event keywords**: Each provider can define keywords and only events tagged with at least one of the keywords will be included. For example, only the `AppStarted` and `DebugMessage` events would be included if we specify the `Startup` keyword. If no keywords are specified, then events with any keyword will be included.

The following is the code that goes into the `Program.cs` to create sample log entries:

```
SampleEventSource.Log.AppStarted("Application Started!");
SampleEventSource.Log.DebugMessage("Process 1");
SampleEventSource.Log.DebugMessage("Process 1 Finished");
SampleEventSource.Log.RequestStart(3);
SampleEventSource.Log.RequestStop(3);
```

Now, you can run your application and use tools such as **PerfView**, **Windows Performance Recorder** (**WPR**), or **Windows Performance Analyzer** (**WPA**) to monitor and analyze the ETW events. For this demo, however, we will use the event viewer built into Visual Studio's diagnostic tools. Open the Performance Profiler in Visual Studio (*Alt + F2*) and select the **Events Viewer** check box. Then, select the small gear icon to the right of the option to open the configuration window. This is shown in *Figure 6.3*.

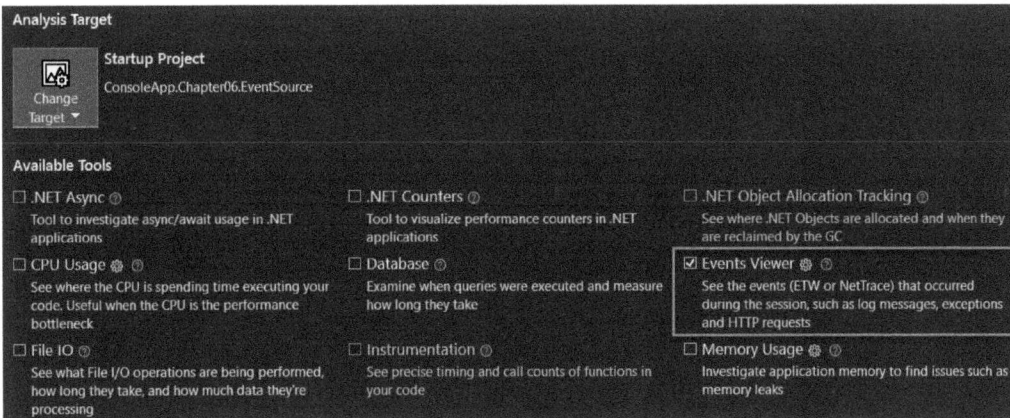

Figure 6.3 – Event Viewer option in Visual Studio's Performance Tools

The new window will contain a table that allows you to specify **Additional Providers**. Proceed to add a row for the SampleEventSource provider, click the **Enabled** checkbox, specify that the **Enabled Keyword** is 0x1, and change the level to **Informational**, as seen in *Figure 6.4*.

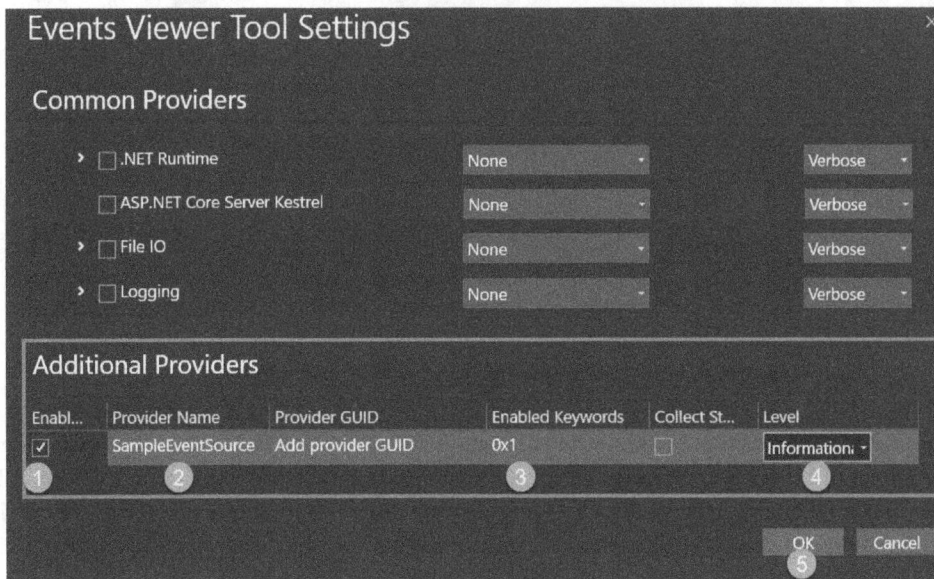

Figure 6.4 – Configuring the additional provider event source for the event viewer

Once all the options have been entered, click **Start** to run the app and collect logs. Select **Stop Collection** or exit the application to stop collecting logs and show the collected data. As seen in *Figure 6.5*, we can filter through the tens of thousands of events generated to view logs generated by our custom event provider.

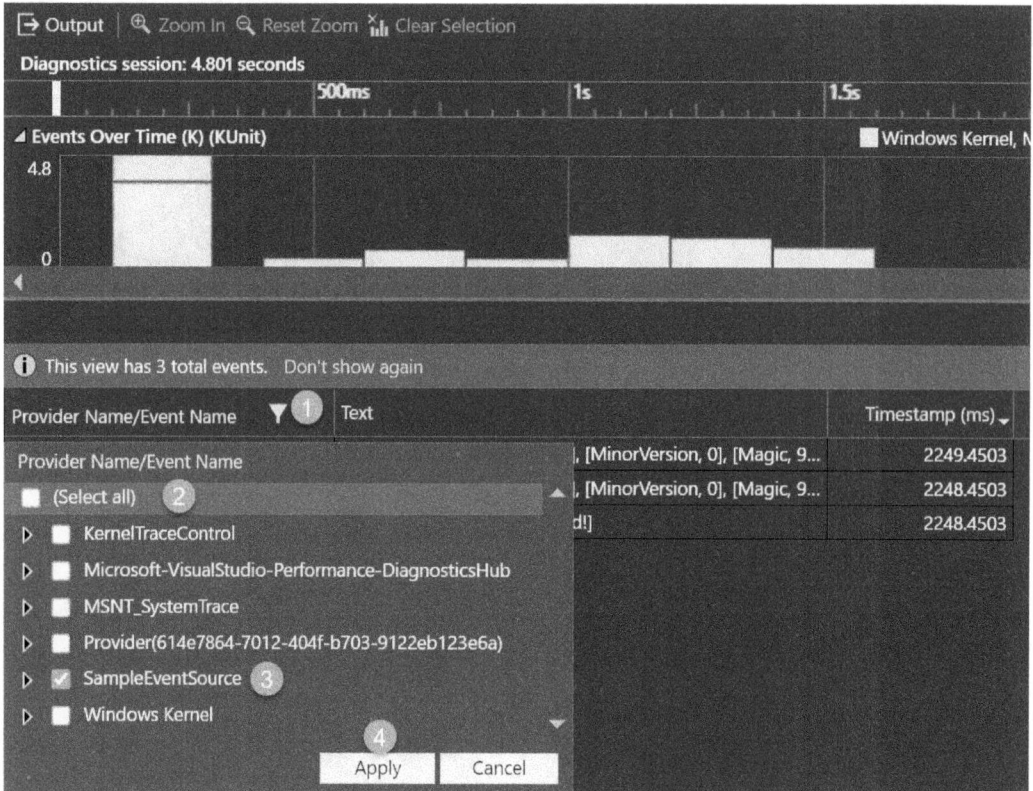

Figure 6.5 – Filter events coming from a custom event provider

This was a simple demo, but it shows how ETW functionality can be embedded into your application. ETW is a powerful logging technology built into many parts of the Windows infrastructure. It is leveraged in the .NET CLR to collect system-wide data and profile all resources (CPU, disk, network, and memory) to help us obtain a holistic view of the application's performance. Given that it has a low overhead, which can be further tuned, it is a suitable solution for monitoring production application diagnostics.

At this point, we have seen several ways to modify both our code and environments to provide additional information about the inner workings of our application at a system and runtime level. These approaches can be beneficial in finding elusive issues but also introduce new challenges for our dev teams. We will discuss some of the downsides to these next.

Downsides of profiling

Memory profiling, while a powerful tool for improving software performance and stability, can introduce performance overhead during operation. This overhead can affect the accuracy and performance of the application being profiled. This concept can be considered a headache, which explains why it is a road less traveled by software developers. Some of the more common challenges that are encountered are as follows:

- **Additional code execution**: Memory profiling tools inject additional code into your application or run alongside it to monitor memory usage. This code tracks every allocation and deallocation, adding extra instructions the processor must execute. The more detailed the profiling (e.g., tracking each memory allocation), the higher the overhead. Also, maintaining and updating logging and instrumentation code can require significant effort, especially as the application grows.

- **Performance overhead**: As the profiling tool tracks memory allocations and deallocations, it needs to store this data somewhere. This involves using additional memory and potentially significant I/O operations to write this data to disk. These operations are not part of the application's normal execution flow and can significantly slow overall performance, especially if the I/O subsystem is already a bottleneck.

- There will also be an increase in CPU usage. Profilers need CPU cycles to run monitoring code, process the collected data, and possibly analyze it on the fly. This additional CPU usage can compete with the application for resources, particularly in CPU-bound scenarios, leading to slower overall performance.

- In addition to tracking the application's memory usage, profilers also require memory to operate. This can include memory for storing the collected data and overhead for the profiler's operations (such as its runtime environment). This increased demand for memory can lead to less available memory for the application, potentially causing more frequent garbage collections or paging, which can degrade performance.

- **Impact on garbage collector**: Profilers can affect the garbage collector's operations. By tracking object allocations and deallocations, the profiler may keep references to objects that would otherwise be collected, thus delaying garbage collection cycles or making them more frequent or prolonged. Each of these scenarios can introduce delays and performance hits to the application.

There are a few strategies that can be employed to mitigate the aforementioned concerns, as follows:

- **Use selective profiling**: Running the profiler only on specific parts of the application or under certain conditions rather than profiling the entire application continuously

- **Off-peak profiling**: Schedule profiling sessions during development or testing phases or during off-peak hours to minimize the impact on production performance

- **Incremental profiling**: Gradually profiling different application parts in successive runs instead of all at once to reduce the load during any single profiling session

Understanding and planning for these overheads is essential when setting up memory profiling, especially in performance-sensitive environments.

Now that we have reviewed some of the challenges and concerns with adding profilers and some ways to manage the potential effects, let's explore how we can detect possible memory leaks using unit testing.

Writing unit tests for memory leaks

As discussed, detection is the first step to addressing a memory leak. Tools such as profilers, debuggers, or heap analyzers, whether built-in or external, can track and analyze memory usage and allocation. These tools help you identify memory leaks by showing how much memory your program uses, where it is allocated, and how it changes over time. Beyond these tools, other formidable testing methods include unit tests.

Unit testing is primarily designed to verify the correctness of code, but it also plays a significant role in memory profiling and optimization. Developers can gain valuable insights into memory usage patterns and potential inefficiencies by integrating memory profiling within unit tests. These tests can simulate and measure memory leaks by running your code under different scenarios, inputs, or loads to check for memory leaks or errors. Unit tests for memory leaks typically involve creating objects, performing operations, and verifying that the objects are correctly garbage collected. Note that detecting memory leaks in unit tests can be challenging and isn't always 100% reliable due to the nature of garbage collection.

Before we start writing unit tests, let us take some time to understand what they are and how they work.

What is unit testing?

Unit testing is a software testing technique where individual units or components of a software application are tested in isolation from the rest of the application. The purpose is to validate that each software unit performs as expected. A unit in this context can be a function, method, procedure, module, or object. It involves writing test cases for all (or as many as possible) methods so that whenever a change causes a fault, it can be quickly identified and fixed. Each test case is executed to check if the unit behaves as expected under various conditions, including edge cases and error conditions.

Each test case should be simple and test a single behavior or aspect of the unit. It should be written to focus on a single functionality, ensuring it does not depend on external systems, such as databases or network services. This allows them to be run multiple times and should produce the same results each time. It also means they can be run automatically, which is beneficial when validations are needed as part of a build process or continuous integration pipeline.

A unit test typically has three steps:

- **Arrange**: This phase prepares the context for executing the method or function being tested. So, initializing objects, setting up dependencies, and preparing test data are all common activities here.

- **Act**: This is where the actual action occurs. The method being tested is invoked, and a result is retrieved accordingly.

- **Assert**: This involves checking the results against the expected outcomes using assertions. If the assertions pass, the test succeeds; if they fail, the test fails.

Now that we have a theoretical understanding of unit testing, let us review how to test for a memory leak.

Testing for a memory leak

First, we create a new solution using the following .NET CLI commands:

```
dotnet new sln -n MemoryLeakSolution
cd MemoryLeakSolution
```

Now that we have the new directory and solution, we will create a console application called **MemoryLeakApp**:

```
dotnet new console -n MemoryLeakApp
```

In this new console application, create a class with a method that leaks memory by not properly managing a list of objects:

```
public class MemoryLeaker;
{
    private List<byte[]> leakingList = new List<byte[]>();

    public void LeakMemory()
    {
        // Each call leaks approximately 10MB of memory.
        byte[] data = new byte[10 * 1024 * 1024]; // Allocate 10 MB
        leakingList.Add(data);
    }

    // Intentionally not freeing the memory
}
```

The `Program.cs` file will contain code to call this class, and the leaky method will be executed several times, like other tests we reviewed in this book:

```
Console.WriteLine("Memory Leak Demo");
var leaker = new MemoryLeaker();
for (int i = 0; i < 10; i++)
{
    leaker.LeakMemory();
    Console.WriteLine($"Iteration {i + 1}: Leaked 10MB");
}
Console.WriteLine("Press any key to exit...");
Console.ReadKey();
```

You must first choose a testing framework to add unit tests to a .NET project. The most used frameworks for .NET are **xUnit**, **NUnit**, and **MSTest**. Each framework has strengths but provides the necessary tools to write and run unit tests. Now, we will create an **xUnit** test to detect the memory leak. We can create a project for this using the following CLI command:

```
dotnet new xunit -n MemoryLeakTests
```

Because these projects need to interact, we will add them both to the solution and as a reference for the `MemoryLeakApp` and the `MemoryLeakTests` projects:

```
dotnet sln add MemoryLeakApp/MemoryLeakApp.csproj
dotnet sln add MemoryLeakTests/MemoryLeakTests.csproj
dotnet add MemoryLeakTests/MemoryLeakTests.csproj reference
MemoryLeakApp/MemoryLeakApp.csproj
```

Now that we have our projects created and referencing each other, we can write the following test:

```
namespace MemoryLeakTests;
public class MemoryLeakTests
{
    [Fact]
    public void TestForMemoryLeak()
    {
        // Arrange
        var leaker = new MemoryLeaker();
        var initialMemory = GetTotalMemoryUsed();

        // Act
        leaker.LeakMemory();

        // Clean up - attempt to force garbage collection for
        // accurate measurement
```

```
        GC.Collect();
        GC.WaitForPendingFinalizers();
        GC.Collect();

        var memoryAfterLeak = GetTotalMemoryUsed();

        // Assert
        Assert.False(memoryAfterLeak > initialMemory,
        "Memory usage increased, leak expected.");
    }

    private long GetTotalMemoryUsed()
    {
        // Forces garbage collection and returns the amount of
        // memory currently allocated in bytes.
        GC.Collect();
        GC.WaitForPendingFinalizers();
        GC.Collect();
        return GC.GetTotalMemory(true);
    }
}
```

The test starts by initializing an instance of MemoryLeaker and recording the initial memory usage. This test will use the System.Diagnostics namespace to measure memory usage before and after the method call. Then, we call the LeakMemory method to simulate the memory leak. After the action, garbage collection is forced to run to get more predictable results, given the challenge that automatic GC might pose based on its natural operation. This will attempt to clean up any unreferenced objects. We then assert that the memory usage has increased post-method call, indicating a memory leak.

To run the test, we run the following commands to build the solution and then run the tests:

```
cd..
dotnet build
dotnet test
```

Since we are testing a method with a leak, the test will fail and present a message like the following:

```
ConsoleApp.Chapter06.MemoryLeakApp.Tests.MemoryLeakTests.
TestForMemoryLeak
  Source: MemoryLeakTests.cs line 6
  Duration: 25 ms

  Message:
Memory usage increased, leak expected.
```

```
    Stack Trace:
MemoryLeakTests.TestForMemoryLeak() line 23
RuntimeMethodHandle.InvokeMethod(Object target, Void** arguments,
Signature sig, Boolean isConstructor)
MethodBaseInvoker.InvokeWithNoArgs(Object obj, BindingFlags
invokeAttr)
```

This outcome helps you verify the presence of a memory leak in the `LeakMemory` method of the `MemoryLeaker` class.

Unit testing is often seen as tedious and cumbersome, but tests help to reduce long-term retesting and help to validate that code operates in a predictable and repeatable way. Although this method of checking for memory leaks requires us to write additional code, it does not require using third-party profiling tools. It can help us detect potential issues before deploying the application into production.

Following best practices for deploying to production is always on the list of priorities since we want that environment to be the most robust and error-free. In the following section, we will review some considerations and good habits that can help us to optimize production deployments.

Production and deployment considerations

At this point, we have reviewed all the best practices and coding methods to ensure that our application is pristine during development. Now, we need to consider getting it into production and into the hands of the very users with whom we want to have a good experience.

Deploying to production is not everyone's favorite activity. It can range from an automated five-minute process to one that takes 12 hours and is still unsuccessful (true story). Production deployments come in all shapes and sizes and, at the root, are relative to the technology, environment, and overall solution being deployed.

Most software deployments might start as manual ones. While straightforward, they have several downsides that can significantly impact the deployment process's reliability, efficiency, and consistency. Here are some of the key disadvantages:

- **Risk of mistakes**: Manual processes are prone to human errors, such as misconfigurations, skipped steps, or incorrect file uploads

- **Consistency issues**: Ensuring consistent deployments across multiple environments (development, staging, production) can be challenging

- **Inefficient**: Manual deployments require significant time and effort, especially for complex applications with numerous components

- **Repetitive tasks**: Many deployment tasks are repetitive and mundane, leading to inefficient use of developer or operations team time

- **Complex rollbacks**: Manually rolling back a deployment in case of failure can be complicated and time-consuming

- **Risk of data loss**: Improperly handled manual rollbacks can result in data loss or corruption, affecting the reliability of the application

- **Slow issue detection**: With manual deployments, detecting issues and getting feedback takes longer, delaying the resolution of problems

For these reasons, having an automated system is a more reliable way to ensure consistency in the deployment process. An added benefit of automated systems is that they can be outfitted with quality checks and gates to ensure that nothing gets deployed unless they meet the minimum standards. This is where **Continuous Integration (CI)** and **Continuous Deployment (CD)** pipelines come into play. Using CI/CD pipelines, we can automate necessary checks, such as unit testing, to ensure the deployed code does not introduce specific issues, including memory leaks. We will review how we can create a CI/CD pipeline next.

Using CI/CD pipelines

In the fast-paced world of software development, CI/CD pipelines have become essential for delivering high-quality software quickly and efficiently. CI frequently integrates code changes into a shared repository, followed by automated builds and tests. CD is the practice of automatically deploying every change that passes the CI process to production. They are usually implemented along with the following components being a part of the workflow:

- **Version control**: Centralized code repositories (e.g., GitHub, GitLab) where code changes are stored

- **Build automation**: Tools and scripts that automate the process of building the application

- **Automated testing**: Integration of unit, integration, and end-to-end tests to ensure code quality

- **Deployment automation**: Scripts and tools that automate the deployment of the application to various environments

One of the more popular and accessible tools for implementing CI/CD pipelines is GitHub. It is already the largest online source control management system and supports hosting source code for most if not all, file and coding framework types. It also has the key **GitHub Actions** feature, which supports the creation of CI/CD workflows that can be easily integrated into your project's repository.

In fact (more about the aforementioned true story), I have used GitHub workflows to reduce a **12-hour manual deployment exercise** for an application spanning five servers, with several web and desktop components, to a **15-minute process** done with a simple commit and no human interaction.

If you don't already have a GitHub account, you can create one for free at www.github.com. To get your code from your computer into a repository, you need to install Git on your PC, which you can retrieve from here https://www.git-scm.com/. *Figure 6.6* shows two options to create a new repository in GitHub.

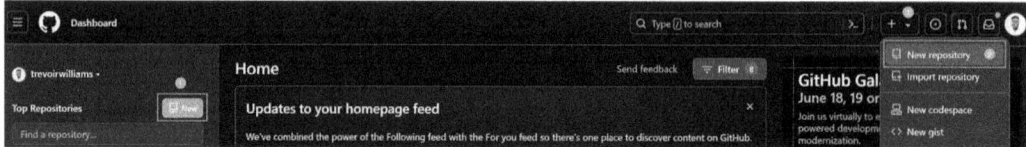

Figure 6.6 – Create a new repository on GitHub

You will be presented with a form, which you can complete by providing a minimum of a **Repository name**. We will use the name MemoryLeakApp for this exercise. This is because we will use the previously created MemoryLeakApp for this exercise.

Run the following CLI commands in the directory of your MemoryLeakApp to initialize the Git repository locally and then push the source code to GitHub:

```
git init
git add .
git commit -m "Initial commit"
git remote add origin https://github.com/trevoirwilliams/
MemoryLeakApp.git
git branch -M main
git push -u origin main
```

Now that our repository is populated with our code, let us add a GitHub action. This action will compile the ASP.NET Core project and then run all the unit tests in the source code.

From the repository window, select the **Actions** option, as highlighted in *Figure 6.7*.

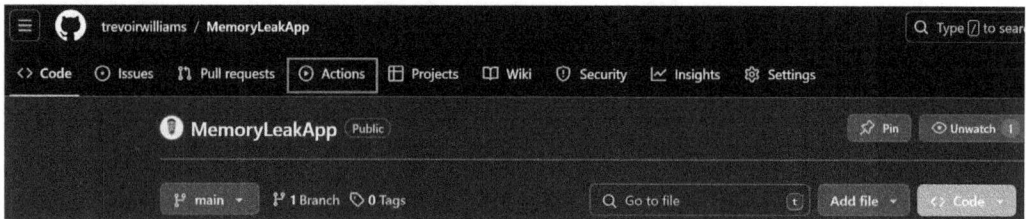

Figure 6.7 – The Actions button for a repository

Scroll down and select the **.NET** template found under the **Continuous Integration** category. See *Figure 6.8* for reference.

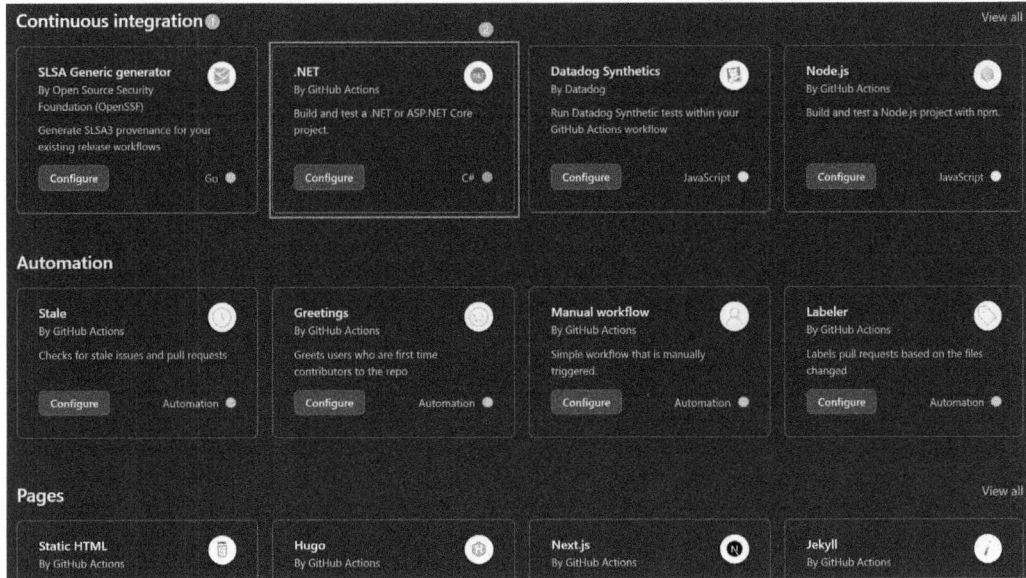

Figure 6.8 – The pre-configured GitHub action for .NET applications

The resulting page is a markdown file that contains .NET compilation instructions. This file is called a **workflow**, which completes several tasks, also called **actions**. The most relevant action, relative to this discussion, is the following.

```
run: dotnet test --no-build --verbosity normal
```

This line will automatically run all the unit tests found in the project. This means that our application with a memory leak will fail the memory leak test, and the build will fail. GitHub will indicate a failed build, and we will use this as a sign that the application is not production-ready. You can add this action to your repository by clicking **Commit Changes**, as seen in *Figure 6.9*.

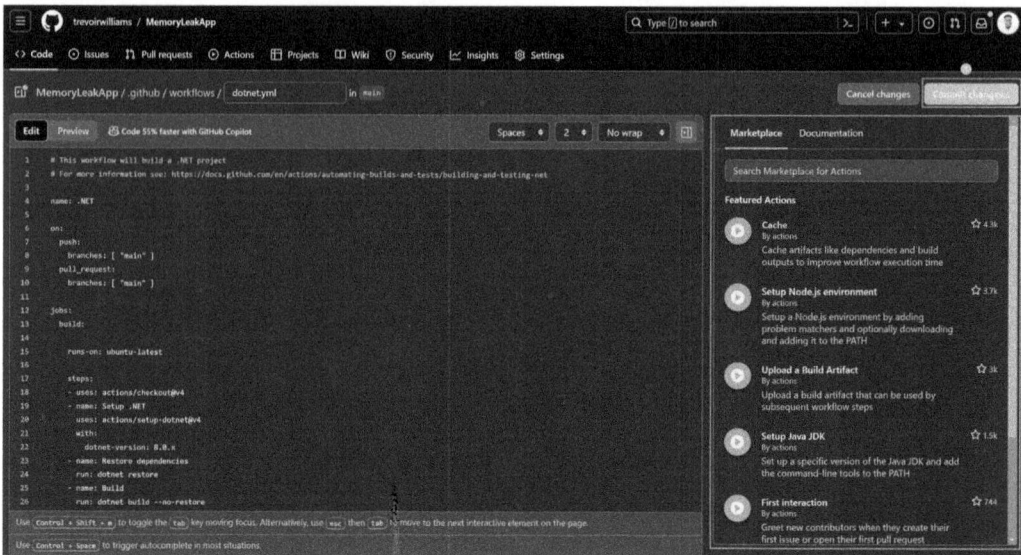

Figure 6.9 – Adding a workflow to your repository

The workflow will be queued for execution immediately and will show a **green tick** for success or a **red x** for failure.

This workflow file can be improved and extended to complete various actions, which can be added to enhance the quality gates and checks you and your organization require during development and before deployments. When editing the workflow file, a panel, as seen in *Figure 6.9*, on the right-hand side, allows you to search the GitHub Marketplace for additional actions. Most actions can be plugged into the existing workflow; some require additional configurations and credentials.

If we are, for instance, deploying to an Azure environment, we can proceed to fund actions that assist us in completing actions such as a load test and the deployment steps. You can automate a load test in Azure Load Testing by creating a CI/CD pipeline to validate your application performance and stability under load continuously.

The following is an example of a workflow that builds an ASP.NET Core web application and has been extended to automatically deploy to an Azure App Service after a successful build:

```
jobs:
  build:
    # ********* shortened for brevity **************

      - name: dotnet publish
        run: dotnet publish -c Release -o ${{env.DOTNET_ROOT}}/myapp

      - name: Upload artifact for deployment job
```

```
      uses: actions/upload-artifact@v2
      with:
        name: .net-app
        path: ${{env.DOTNET_ROOT}}/myapp

  deploy:
    runs-on: windows-latest
    needs: build
    environment:
      name: 'Production'
      url: ${{ steps.deploy-to-webapp.outputs.webapp-url }}

    steps:
      - name: Download artifact from build job
        uses: actions/download-artifact@v2
        with:
          name: .net-app

      - name: Deploy to Azure Web App
        id: deploy-to-webapp
        uses: azure/webapps-deploy@v2
        with:
          app-name: 'azure-app'
          slot-name: 'Production'
          publish-profile: ${{ secrets.AZUREAPPSERVICE_
          PUBLISHPROFILE_7D7FA20F3A9D4B3C8C90F1DF44E61B14 }}
          package: .
```

The application will be packaged after the build jobs have been completed successfully. Next, the deploy job will run and proceed to unwrap the package from the build job, connect to the target Azure application service, and then deploy. GitHub has robust documentation on GitHub actions and possibilities, and you can explore further scenarios here: `https://docs.github.com/en/actions/quickstart`.

We have automated deployment using a GitHub workflow while ensuring that minimum requirements are met simultaneously. Now that we know how to check our application before we deploy it, let us review how we can monitor an application after it has been deployed in a cloud-based environment.

Monitoring cloud environments

We have focused on why memory management is critical in application development. While a local machine might provide an environment where a memory leak might lead to temporary downtime and require triaging, a cloud environment also introduces higher costs if the application is not optimized for the environment.

As memory usage increases due to leaks, the application may experience slow response times, affecting the overall performance and user experience. Since memory leaks can cause increased garbage collection activity, this will increase latency as the garbage collector tries to free up memory.

Cloud resources are usually provisioned based on specific resource availability. This means an application needs to be optimized to operate in an environment that might provide limited resources upfront. If a memory leak is present, it can lead to frequent crashes and downtime, affecting the availability of services and potentially breaching **service-level agreements** (**SLAs**). This might also lead to over-provisioned resources since you may pay more for resources based on what you think is required for the application. Even more so, if you configure auto-scaling based on resource usage, an application with a memory leak might require additional instances of the hosting resources (for example, **virtual machines** (**VMs**)) to handle the increased memory usage.

Hopefully, you will catch the memory leak before it gets to this point and your users start sending rude emails.

There are several cloud providers, and there are several tools. Some tools are provided directly by the cloud providers, and others are platform agnostic. Most, if not all, are equipped with robust monitoring and instrumentation features and can trigger alerts if certain thresholds are crossed. Some popular ones are the following:

- Azure Monitor Application Insights: A feature of Azure Monitor that excels in **Application Performance Management** (**APM**) for live web applications.

- AWS CloudWatch: A control center that monitors all your AWS resources in one place.

- Stackdriver Profiler: A statistical, low-overhead profiler that continuously gathers CPU usage and memory-allocation information from your production applications. Used by **Google Cloud Platform** (**GCP**).

- Site24x7: Monitors the health and performance of websites, servers, networks, applications, and cloud platforms.

Since we are demonstrating deployments to **Microsoft Azure**, we will continue with examples for this environment. **Azure Monitor Application Insights** is a great tool to add to your Microsoft Azure environment.

This service offers several features, but the most useful ones within the context are the following:

- **Application dashboard**: An at-a-glance assessment of your application's health and performance

- **Application map**: A diagram overview of application architecture and components' interactions

- **Live metrics**: A real-time analytics dashboard for insight into application activity and performance

- **Alerts**: Monitor various aspects of your application and trigger various actions

- **Metrics**: Dive deep into metrics data to understand usage patterns and trends

This service is a part of the **Azure Monitor** offering and provides **Application Performance Monitoring (APM)** features. APM tools are helpful for monitoring applications from development, through testing, and into production by proactively understanding how an application is performing and reactively reviewing application execution data to determine the cause of an incident.

Application Insights can be added to your .NET application with a few easy steps. First, you can add the `Microsoft.ApplicationInsights.AspNetCore` package using the following command:

```
dotnet add package Microsoft.ApplicationInsights.AspNetCore
```

Next, in the `Program.cs` file of the application, you need to add the following line:

```
builder.Services.AddApplicationInsightsTelemetry();
```

Next, you must add a configuration to your `appsettings.json` file, where you place the connection string to the Application Insights service:

```
{
  ...
  "ApplicationInsights": {
    "ConnectionString": "Copy connection string from Application
    Insights Resource Overview"
  }
}
```

This assumes that the service has already been created and enabled. When you run your application and make requests to it, telemetry and instrumentation data will now flow to Application Insights. To learn more about how Application Insights and monitoring can be used, reference the excellent documentation here: `https://learn.microsoft.com/en-us/azure/azure-monitor/app/app-insights-overview`.

Now that we have explored several ways of ensuring the quality of the application being deployed to production and how we can monitor cloud environments, let's wrap up this chapter.

Summary

This chapter delved into the critical aspects of memory profiling and optimization in .NET development, equipping developers with the knowledge and tools to identify, analyze, and mitigate memory-related issues. We started by reviewing the fundamentals of memory profiling, which is the process of monitoring and analyzing an application's memory usage to identify potential inefficiencies, leaks, and bottlenecks. We also reviewed some commonly used tools that help to analyze system resource usage and give reports and insights.

We then explored several practical approaches to profiling, which generally required authoring additional code and using third-party tools. Even though profiling is a valuable and necessary measure to keep our applications as efficient as possible, it has several downsides. Third-party tools increase system resource usage, which can lead to skewed results, and the additional code that needs to be written might also deter developers from exploring this concept.

We also looked at low-impact options including Event Tracing for Windows, which allows developers to define specific log messages with details on system operations during runtime. This is a great option, but it is limited to Windows-based machines.

We then looked at an option that is tried, true, and tested, allowing us to detect potential memory leaks without needing a third-party analysis tool. This is unit testing, which is a widely accepted programming practice. This enables us to collect memory usage statistics before and after we test a method and then assume that the memory usage is the same. If it is not, then we have a potential leak. Unit tests can be integrated into automated processes and deployments, allowing us to spot issues in our code before deployment.

Finally, we explored some nuances surrounding deploying to production and reviewed the efficacy of implementing automated pipelines to introduce consistency and additional quality gates during development and before deployments. We reviewed how GitHub Actions can facilitate these checks and complete the deployment. We then reviewed some basic steps towards adding Azure monitoring to a web application so that we could observe abnormal memory usage patterns in the deployed application.

In the next chapter, we will explore low-level programming in memory management.

7

Low-Level Programming

High-level programming languages such as C# have revolutionized software development by offering a rich set of functions and abstractions that simplify development tasks, enhancing productivity and code maintainability. However, as powerful as these abstractions are, there are scenarios where more deliberate control over the system is required. These scenarios require understanding low-level programming techniques and can make up the difference between high-performance and slower applications.

Low-level programming refers to coding practices that operate closer to the hardware, allowing the developer to have more direct control of memory and system resources. In contrast to high-level programming, which abstracts away most of the complexity involved in these operations, low-level programming requires a detailed understanding of the underlying architecture, as well as memory management and performance optimization techniques.

This chapter will explore the essential concepts and techniques of low-level programming in .NET. We will review concepts such as the following:

- Working with unsafe code
- Allocating and deallocating unmanaged memory
- Interoperability with unmanaged code

By the end of this chapter, you will have a solid foundation in low-level programming within the .NET ecosystem, equipped with the knowledge and skills to enhance your applications' performance and capabilities. Whether you're optimizing existing code or developing new solutions, mastering these techniques will enable you to push the boundaries of what your software can achieve.

Let us begin by discussing unmanaged memory and how allocation and deallocation operations are handled.

Technical requirements

- You will need to have access to Visual Studio 2022, which can be found at `https://visualstudio.microsoft.com/vs/community/`.

- You will also need Visual Studio Code, which you'll find at `https://code.visualstudio.com/`.

- Lastly, you'll need .NET 8 SDK, which you can find at `https://dotnet.microsoft.com/en-us/download/visual-studio-sdks`.

Working with unsafe code

The `unsafe` keyword denotes a section of code that is not managed by the **Common Language Runtime** (**CLR**) or by unmanaged code. `Unsafe` is used to declare a type or member or specify a block code. When used to qualify a method, the context of the entire method is unsafe.

We will mention managed and unmanaged code several times while discussing low-level programming and unsafe code. As a reminder, managed code executes under the supervision of the CLR and the **Garbage Collector** (**GC**). They perform housekeeping tasks such as the following:

- Managing memory for objects

- Performing type verification

- Doing garbage collection

Managed code in .NET is generally considered **verifiably safe code**, meaning that the .NET development tools can verify that the code is safe. The primary attribute of safe code is that it doesn't directly access memory using pointers, allocate raw memory, or create managed objects.

On the other hand, unmanaged code executes outside the supervision of the CLR. In this case, the developer is responsible for the following:

- Calling the memory allocation method

- Ensuring proper type casting

- Ensuring that memory is released when the process is completed

Note that there is a difference between *unsafe* and *unmanaged* code. Unmanaged code runs outside the context of CLR and the supervision of the GC. Unsafe code, however, runs under the context of CLR but allows for the use of pointers for direct memory access. It is code that exists outside of the verifiable subset of **Common Intermediate Language** (**CIL**). This is called the **unsafe context**.

C# supports an unsafe context wherein unverifiable code can be authored. In this context, a developer can perform otherwise restricted actions such as using pointers, allocating and freeing memory blocks, and calling methods using function pointers. Unsafe code isn't necessarily dangerous; it is just code that cannot be verified for safety. Some properties of unsafe code include the following:

- It may increase performance by removing array bounds checks

- It is required to call native functions that require pointers

- It introduces potential security and stability risks

- The code containing unsafe blocks must be compiled with the **AllowUnsafeBlocks** compiler option

An unsafe context is created by using the unsafe modifier in the declaration of a type, member, or local function, or by using an unsafe statement:

- Adding the unsafe modifier to a class, struct, interface, or delegate declaration makes the entire type declaration and its body an unsafe context. If the type declaration is partial, only that part is an unsafe context.

- Adding the unsafe modifier to a field, method, property, event, indexer, operator, instance constructor, finalizer, static constructor, or local function makes the entire member declaration an unsafe context.

- An unsafe statement allows using an unsafe context within a block, making the entire block an unsafe context. A local function declared within an unsafe context is also considered unsafe.

The following code snippet shows an example of an unsafe method being implemented and called in a C# program:

```
static unsafe void UnsafeMethod()
{
    int num = 10;
    int* p = & num;

    Console.WriteLine("Value: {0}", num);
    Console.WriteLine("Address: {0:X}", (int)p);
}

UnsafeMethod();
```

If you attempt to run this code in Visual Studio, you may get the result shown in *Figure 7.1*.

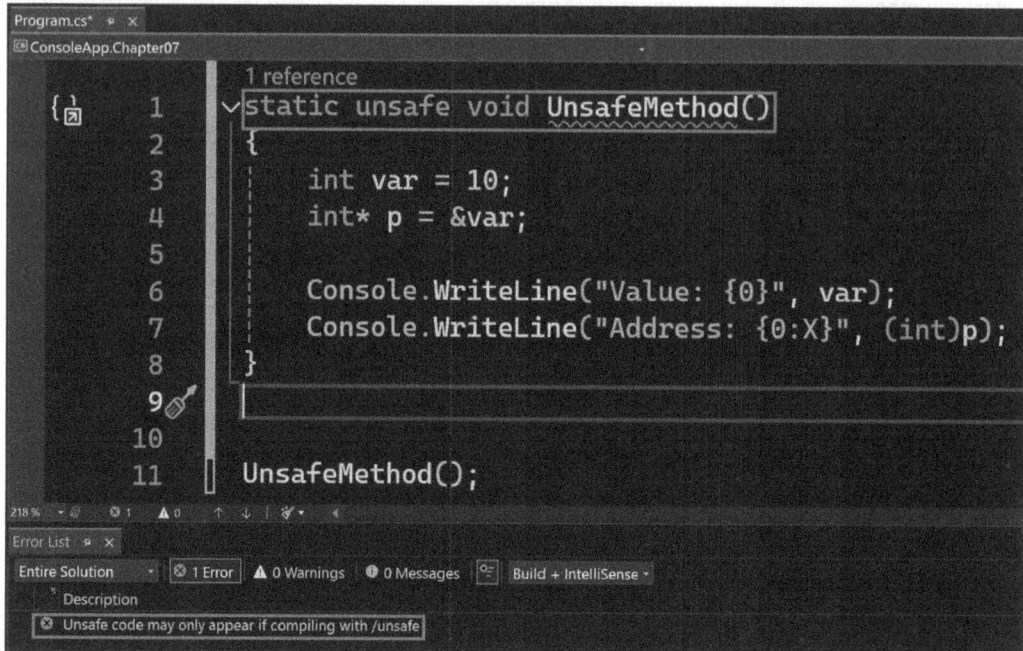

Figure 7.1 – Attempting to run unsafe code in Visual Studio

To fix this issue, you need to enable the unsafe context in Visual Studio by following these instructions:

1. In Visual Studio, right-click on the project in **Solution Explorer** and select **Properties**.
2. Go to the **Build | General** tab.
3. Check the **Allow code that uses the 'unsafe' keyword to compile** option.

The preceding steps are shown in *Figure 7.2*.

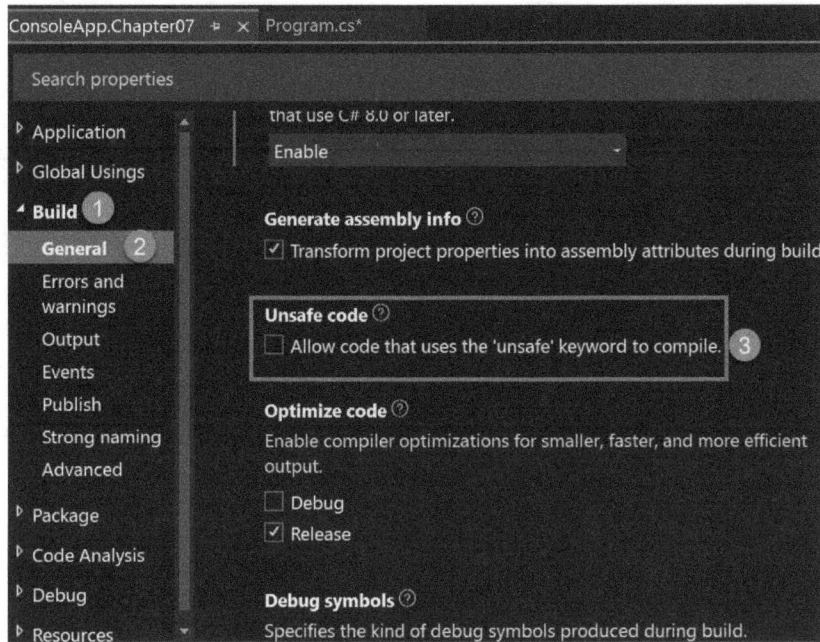

Figure 7.2 – Enabling unsafe code in a Visual Studio project

Alternatively, you can enable **AllowUnsafeBlocks** by editing the `csproj` file of the target project using the following code. This approach works regardless of the IDE:

```
<Project Sdk="Microsoft.NET.Sdk">
    <PropertyGroup>
        <AllowUnsafeBlocks>True</AllowUnsafeBlocks>
    </PropertyGroup>
</Project>
```

Once this is enabled, the error will disappear and the application will execute, as seen in *Figure 7.3*.

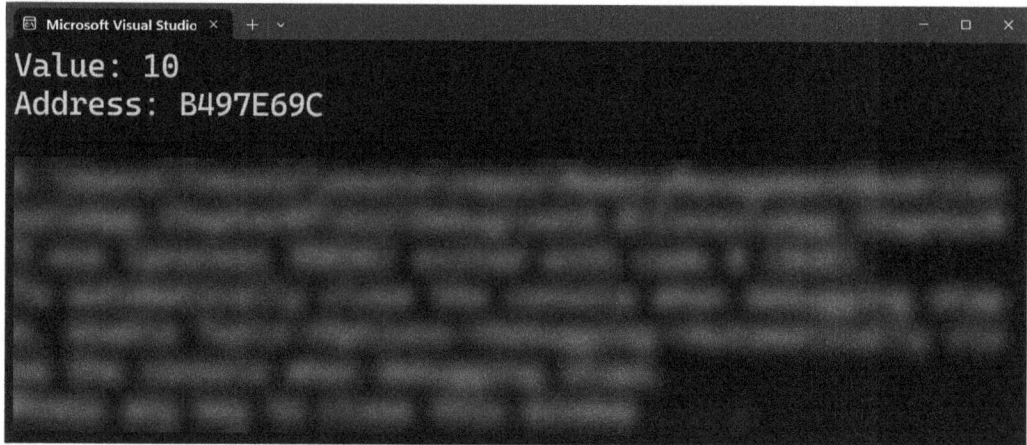

Figure 7.3 – Unsafe application execution

One of the recurring topics here is working with pointers in C#. Let's examine how we can do this with unsafe code.

Pointers and pointer operations

A pointer is a variable that is designed to hold the memory address of another object. Essentially, it points to a specific memory location. A pointer must share the same data type as the variable it references. Therefore, for a pointer to refer to an `integer` variable's memory location, it must be of the `integer` type. When you create a pointer, you instruct the compiler to allocate a variable that will store the address of another variable in memory. Pointers are crucial in C# when working with unsafe code, as they can access and modify the memory addresses of other variables.

By using pointers, you can safely access and alter the memory addresses of variables. Pointers facilitate access to the contents of a memory location and are used for passing data between functions, manipulating data structures, and implementing linked lists. Pointers can point to only unmanaged types, including all basic data types, enum types, other pointers, and structs that contain only unmanaged types.

Some examples of using unmanaged types to define pointers are as follows:

- `byte*`: Pointer to `byte`
- `char*`: Pointer to `char`
- `int**`: Pointer to pointer to `int`
- `int* []`: Array of pointers to `int`
- `void*`: Pointer to unknown type

The `void*` pointer type can be used when the referent type is unknown. A pointer of this type cannot use the indirection operator, nor can arithmetic be performed on such a pointer. However, this pointer type can be cast to and from any other pointer type and compared to values of different pointer types. A pointer is defined using the following syntax:

```
type* variable_name;
```

The type specified before the `*` is called the **referent type**. Only unmanaged types can be used as the `referent` type. The `*` acts as a dereference or indirection operator, which syntactically means that we intend to access the value stored at the memory address. Following the code example from the previous section, we defined an integer variable and then a pointer to the address of this variable using the following code:

```
int num = 10;
int* p = &num;
```

The `&` operator is used to obtain the memory address of the variable number, and the `*` operator is used to access the value stored at the memory address. An example of a memory address is the output of the p value, as seen in *Figure 7.3*. Recall from the earlier chapters that memory is organized as a sequence of bytes with unique addresses. A memory address is essentially an index in this sequence. Variables are allocated a block of memory to store a value. The address of this block can be accessed using pointers.

Dereferencing a pointer means accessing the value stored at the memory address the pointer is pointing to, as seen in the following code:

```
int value = *p;   // value now holds 10
```

It is noteworthy that pointers are not tracked by the GC. For this reason, pointers cannot point to a reference type (standard C# object) or a struct containing reference types. Since pointers are unmanaged, a pointer type may also be used as the referent type for another pointer.

The following is a list of operators and statements that can operate on pointers in an unsafe context:

- `*`: Indirection operator
- `->`: Used to access a struct member through a pointer
- `[]`: May be used to index a pointer
- `&`: May be used to get a variable's address
- `++` and `--`: Used to increment and decrement pointers
- `+` and `-`: Binary operators that may be used to perform pointer arithmetic
- `==`, `!=`, `<`, `>`, `<=`, and `>=`: May be used to compare pointers

- `stackalloc`: A statement used to allocate memory from the call stack

- `fixed`: A statement used to fix a variable so its address can be obtained temporarily

The code that follows shows pointer arithmetic operations, such as addition and subtraction, while iterating through elements of an array:

```
unsafe {
    int[] numbers = { 10, 20, 30, 40 };
    fixed (int* ptr = numbers) {
        for (int i = 0; i < numbers.Length; i++) {
            Console.WriteLine(*(ptr + i));
            // Outputs each element in the array
        }
    }
}
```

In this example, `ptr + i` moves the pointer to the **ith** element of the array.

These are simple pointer examples using straightforward data types. Recall that pointers can also be assigned a struct type, which, while unmanaged, allows us to define a more complex variable structure. At this point, we need to access the struct members using the `->` operator.

Before diving into pointer member access, let us review why classes and structs are different:

- **Structs**: Value (unmanaged) types allocated on the stack and passed by value

- **Classes**: Reference (managed) types allocated on the heap and passed by reference

Pointer member access primarily applies to structs, as they are contiguous in memory, making calculating offsets for their fields easier. Consider the following code example:

```
public struct Point {
    public int X;
    public int Y;
}

unsafe {
    Point point = new Point { X = 10, Y = 20 };
    Point* ptr = &point;

    // Using the -> operator
    Console.WriteLine(ptr->X); // Outputs 10
    Console.WriteLine(ptr->Y); // Outputs 20

    // Using the * and . operators
```

```
        Console.WriteLine((*ptr).X); // Outputs 10
        Console.WriteLine((*ptr).Y); // Outputs 20
}
```

Here, we define a struct named `Point`, with two public integer fields: X and Y. Structs in C# are value types, which means that they are typically stored on the stack. We already know that we need the `unsafe` keyword to enable pointer operations. In this unsafe block, we initialize the `Point` struct and define a `ptr` pointer to it. The & operator is used to get the memory address of the `point` variable.

The `->` operator is used to access the struct members to which the `ptr` pointer points. `ptr->X` and `ptr->Y` access the X and Y fields of the `Point` struct at the memory address stored in `ptr`. The values 10 and 20, respectively, are printed to the console.

When the * operator is used to dereference the `ptr` pointer, it effectively gets the value at the memory address it points to. After dereferencing, the dot operator (.) is used to access the struct members. `(*ptr).X` and `(*ptr).Y` access the X and Y fields of the `Point` struct.

Importantly, pointer validation is crucial to ensure that a pointer is not null before it is dereferenced. Dereferencing a null pointer leads to undefined behavior and can cause the application to crash. This validation step can be a simple null check with an `if` statement, as depicted in the code that follows:

```
if (ptr == null)
{
    Console.WriteLine("Pointer is null. Cannot dereference.");
}else{
    // Perform pointer related operations
}
```

Pointers are potent constructs for developing performance-critical applications, as well as for systems programming or interfacing with unmanaged code. However, they require a deep understanding of memory management and careful handling to avoid common pitfalls. Here are some best practices to consider:

- Always validate pointers before dereferencing them to avoid getting null pointer exceptions or accessing invalid memory locations.

- Properly manage the memory allocation and deallocation operations using `stackalloc`, the `Marshall` class, or the `SafeHandle` class.

- Minimize the use of unsafe code to the smallest possible scope. Encapsulate unsafe operations within methods or classes to reduce the risk of errors.

- Always use the `fixed` statement to pin objects in memory and prevent the GC from moving them when dealing with arrays or fixed-size buffers.

- Be aware of your structures' and classes' memory layout. Getting this wrong can lead to incorrect pointer arithmetic and data corruption.

- Thoroughly test and debug code that involves pointers. Use debugging tools to inspect memory and validate pointer operations.

Now that we understand how to implement pointers in our code, let's examine the `fixed` statement and how it helps us handle unmanaged memory.

Using the fixed statement

The `fixed` statement is a critical tool for safely working with unmanaged memory and pointers. It prevents the GC from relocating a managed object and ensures that pointers to the object's memory remain valid. The `fixed` statement pins a managed object in memory, preventing the GC from moving it. In the preceding examples, wherein local variables were defined, we could rest assured that the variables would exist on the stack, not the heap. Variables in storage locations affected by the GC are called **movable variables**. Object fields and array elements are examples of movable variables. Their addresses can be retrieved if they are fixed with a `fixed` statement. The obtained address is valid only inside the block of a `fixed` statement.

The following example shows how to use a `fixed` statement and the & operator:

```
fixed (type* pointer = &managedVariable) {
    // Code that uses the pointer
}
```

`type* pointer` declares a pointer of the specified type and `&managedVariable` obtains the address of the managed variable. The following code snippet shows how the fixed statement can be used for an array. Arrays in .NET are managed objects; pinning them ensures that their elements remain in a fixed memory location:

```
int[] numbers = { 1, 2, 3, 4, 5 };

unsafe
{
    fixed (int* ptr = numbers)
    {
        for (int i = 0; i < numbers.Length; i++)
        {
            Console.WriteLine(ptr[i]);
        }
    }
}
```

Now that the array is pinned in memory, the GC will not collect the `numbers array` object. This ensures that the `ptr` pointer can access the elements without the risk of a null reference error.

Another common scenario for pinning is with strings, which are immutable managed objects. The fixed statement can pin strings when passing them to unmanaged code that requires a stable memory address:

```
string text = "Hello, World!";

unsafe
{
    fixed (char* ptr = text)
    {
        for (int i = 0; i < text.Length; i++)
        {
            Console.WriteLine(ptr[i]);
        }
    }
}
```

In some cases, you may need to pin multiple objects simultaneously. This can be achieved by nesting `fixed` statements:

```
int[] array1 = { 1, 2, 3 };
int[] array2 = { 4, 5, 6 };

unsafe
{
    fixed (int* ptr1 = array1)
    fixed (int* ptr2 = array2)
    {
        for (int i = 0; i < array1.Length; i++)
        {
            Console.WriteLine($"Array1[{i}] =
            {ptr1[i]}, Array2[{i}] = {ptr2[i]}");
        }
    }
}
```

Here, we pin both `array1` and `array2` simultaneously and allow the `ptr1` and `ptr2` pointers to access and print the elements of both arrays.

While the fixed statement is powerful, it should be used judiciously:

- Limit the scope of the fixed statement to the smallest possible block of code. This minimizes the duration for which objects are pinned, reducing the impact on the GC.

- Pinning large objects can fragment the managed heap. Consider pinning only necessary portions of data.

- Ensure that the fixed block handles exceptions correctly. This prevents pinned objects from remaining pinned longer than necessary.

- The `fixed` statement is handy when interoperating with unmanaged code via `P/Invoke`. Pin objects before passing their pointers to unmanaged functions to ensure that their memory locations remain stable.

Next, we will look at allocating memory using `stackalloc` and `Marshall` as the main ways to develop an application using unsafe code.

Allocating and deallocating unmanaged memory

As we have seen since *Chapter 1*, .NET runtime handles most memory management tasks through the built-in GC, freeing developers from the chores of manual memory allocation and deallocation. Managed memory offers several advantages, including ease of use, safety from common errors such as buffer overflows, and reduced memory leaks. While the benefits of relying on the automatic management of memory are clear, in this chapter, we are taking a step back to explore when more than reliance on this mechanism might be required. While the GC is convenient, it can lead to inefficiencies and performance issues.

Recall that the GC is responsible for automatic memory management in .NET. It tracks object references, identifies unused objects, and reclaims memory. The GC operates in multiple generations:

- **Generation 0**: Short-lived objects

- **Generation 1**: Objects that survive a Generation 0 collection

- **Generation 2**: Long-lived objects

The GC runs periodically and without the explicit intervention of a developer, making it less predictable than sometimes desired. While it dramatically simplifies memory management, it sometimes risks introducing performance overhead, since more frequent collection operations can cause noticeable delays in performance-critical applications such as real-time data processing or gaming.

Unmanaged memory is allocated and managed outside the control of the .NET GC. It is typically allocated using lower-level APIs and must be manually handled by the developer. This is referred to as low-level programming.

Low-level programming often involves the following:

- **Direct memory manipulation**: This is the process of using pointers and unsafe code to manipulate memory directly. Low-level programming offers the necessary tools to manage resources efficiently and predictably.

- **Interoperability with unmanaged code**: This is the process of calling and interacting with functions from unmanaged libraries, such as Windows API, using **Platform Invocation Services (P/Invoke)**.

- **Custom memory management**: This describes allocating and deallocating memory manually to optimize resource usage and application performance.

- **Performance optimization**: This term refers to fine-tuning code to reduce overhead and improve execution speed by efficiently minimizing garbage collection and managing large objects.

- **Learning and mastery**: This involves diving into low-level programming to enhance your understanding of how the .NET runtime and underlying hardware work. This helps you write more efficient code and debug complex issues.

The most-touted reason for implementing low-level programming and unmanaged memory allocation techniques in C# is that it gives developers fine-grained control over memory. Still, it comes with significant risks and challenges. Here are some of the primary dangers associated with unmanaged allocation:

- **Memory leaks**: Memory leaks occur when memory has been allocated and is not freed correctly. The risk of this happening significantly increases with unmanaged allocation since the developer must be far more deliberate and consistent in code to handle unmanaged allocations.

- **Dangling pointers**: A dangling pointer refers to a pointer that still references a memory location after it has been freed. Dereferencing a dangling pointer can lead to undefined behavior, including application crashes or data corruption.

- **Buffer overflows and underflows**: These can occur when writing more data to a buffer than it can hold, potentially overwriting adjacent memory. Buffer underflows happen when reading before the start of a buffer. Both can lead to data corruption, security vulnerabilities, and crashes.

- **Memory fragmentation**: Frequent allocation and deallocation of unmanaged memory can lead to fragmentation, whereby free memory is divided into small, non-contiguous blocks. This can make it difficult to allocate large memory blocks and reduce overall memory utilization efficiency.

- **Complexity and maintenance overhead**: Manual memory management increases code complexity, as well as the maintenance burden. It requires careful tracking of memory allocations and deallocations, increasing the risk of bugs and making the code harder to maintain and understand.

- **Thread safety issues**: When multiple threads access unmanaged memory, race conditions and other concurrency issues can occur, leading to data corruption and unpredictable behavior.

Low-level programming and manual memory management provide precise control, reducing latency and improving responsiveness, which offers better control and optimization capabilities in certain scenarios. Low-level programming techniques allow developers to allocate memory on the stack or use custom memory pools, reducing the frequency of garbage collections.

C# allows us to use several techniques and keywords to orchestrate manual allocation and deallocation activities. The first one we will explore is `stackalloc`.

Using the stackalloc keyword for stack allocation

`stackalloc` is a C# keyword that allows memory to be allocated on the stack instead of the heap. Recall that the .NET CLR boasts an area called the managed heap, which is managed automatically by the GC to handle objects' life cycles. This automation simplifies memory management for developers, ensuring that unused objects are eventually cleaned up without requiring manual intervention. The general stages of allocations to the managed heap are as follows:

1. **Object creation**: Memory is allocated on the heap when an object is created using the new keyword.

2. **Reference tracking**: The .NET runtime maintains references to objects in the heap, tracking their usage.

3. **Garbage collection**: The GC periodically scans the heap to identify objects that are no longer referenced by the application. These objects are marked for collection and their memory is reclaimed.

We have discussed all the pros and cons of the heap up until now, and we are aware of the general behaviors of heap allocation, such as the following:

*   It allocates memory on the managed heap

*   The GC manages memory

*   It is safe and does not require an unsafe context

*   It has slower allocation and deallocation due to garbage collection

*   It is suitable for larger and longer-lived allocations

Any time we use the new keyword in C#, we can be sure that a heap allocation will take place in memory for that object. What follows is a code snippet showing a standard example of a new array object being declared to the heap:

```
public void ProcessDataHeap()
{
    const int bufferSize = 1024;
    byte[] buffer = new byte[bufferSize];
    // Allocate 1024 bytes on the heap

    // Process data into the buffer
    for (int i = 0; i < bufferSize; i++)
    {
        buffer[i] = (byte)i;
    }

    // Buffer will be freed by the garbage collector eventually
}
```

This is the code that we have become accustomed to seeing when declaring a new object. `byte []` `buffer = new byte[bufferSize]` allocates a block of memory on the heap and the `new` keyword creates a new array. The GC will eventually free the memory allocated on the heap when it determines that the buffer is no longer in use. Some of the implications of what happens in memory are as follows:

- **Slower allocation**: Allocating memory on the heap involves more complex operations

- **Runtime pauses:** The GC periodically runs to free unused memory, which can cause unpredictable pauses

- **Longer lifetime:** Objects allocated on the heap can persist beyond the method's scope if there are references to them

Stack allocation is much faster than heap allocation. This is because allocating memory on the stack is a simple operation that involves moving the stack pointer, while heap allocation requires managing the heap's data structures. When memory is allocated with `stackalloc`, the GC does not manage it. This means it is automatically reclaimed when the method scope ends, reducing the frequency and impact of garbage collection pauses. *Figure 7.4* depicts this scenario's general steps between memory allocation and deallocation.

Memory Allocation with stackalloc in .NET Programming

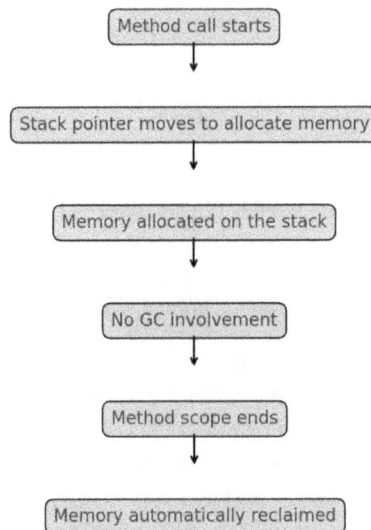

```
                    Method call starts
                           |
                           v
          Stack pointer moves to allocate memory
                           |
                           v
              Memory allocated on the stack
                           |
                           v
                   No GC involvement
                           |
                           v
                   Method scope ends
                           |
                           v
            Memory automatically reclaimed
```

Figure 7.4 – Memory is allocated on the stack and managed without the GC

This allocation method also ensures that memory is allocated contiguously, leading to better performance, especially for applications that process data sequentially.

`stackalloc` is best used in scenarios where temporary memory that is limited in size is needed and the benefits of fast allocation and deallocation outweigh the limitations of stack-based memory. Typical use cases include the following:

- **Performance-critical Ccode**: Such algorithms require temporary buffers for quick computations

- **Real-time systems**: Where predictability and low-latency memory operations are crucial

- **Fixed-size buffers**: When working with fixed-size data structures such as small arrays or structs within a known scope

Following the previous example where a byte array was declared and allocated to the managed heap, we can limit its allocation to the stack using the following refactor with `stackalloc`:

```
public unsafe void ProcessData()
{
    const int bufferSize = 1024;
    byte* buffer = stackalloc byte[bufferSize];
    // Allocate 1024 bytes on the stack

    // Process data into the buffer
    for (int i = 0; i < bufferSize; i++)
    {
        buffer[i] = (byte)i;
    }

    // Buffer is automatically freed when the method exits
}
```

The `unsafe` keyword allows for pointer manipulation in C# and is required for direct memory access operations. `byte* buffer = stackalloc byte[bufferSize]` allocates a block of memory on the stack, and `stackalloc` allocates memory directly on the stack frame of the current method. The memory is automatically reclaimed when the method exits, leading to automatic memory deallocation.

The `unsafe` keyword enables direct access to and modification of memory. Since the GC, by default, ensures type safety and guarantees automated memory management, we can bypass the CLR's automated memory management methods by using this keyword. `unsafe` is very powerful; remember that **with great power comes great responsibility**. It should be used when you need absolute control over an operation's allocation operation and only under specific circumstances. Directly manipulating memory can lead to many issues, including memory leaks, buffer overflows, and other security vulnerabilities, as previously mentioned. Generally, you should only use `unsafe` if you fully understand the risks and benefits of this feature. If you are not confident in your resolve, then allow the CLR and GC to do what they do best.

Stack allocation is much faster than heap allocation. This is because allocating memory on the stack is a simple operation that involves moving the stack pointer, while heap allocation requires managing the heap's data structures. The GC will also have no custody over memory allocated with `stackalloc`, reducing the frequency and impact of garbage collection pauses.

Now, while `stackalloc` provides several performance benefits, it does have several limitations that need to be explored:

- **Stack size limitation**: The stack size is limited and excessive use of `stackalloc` can lead to a stack overflow. Use it for small to moderate allocation operations.

- **Unsafe context**: `stackalloc` requires an unsafe context, which means that the code must be marked with the `unsafe` keyword. This requires additional caution and security considerations.

- **Scope bound**: The lifetime of the allocated memory is bound to the method scope, making it unsuitable for data that needs to persist beyond the method execution.

In summary, `stackalloc` is a powerful tool for situations wherein short-lived objects must be created and performance is highly critical. Be mindful of the potential pitfalls of this kind of operation and, even more importantly, the limitations of the supporting stack space needed for these allocations. The stack allocation and heap allocation trade-offs must be carefully considered when writing code. Given the fact that using `stackalloc` will limit the lifetime of an object to the method within which it is declared, we must consider a scenario wherein we would like the object to persist beyond a single method while not existing on the heap. We will use the `Marshal` class and we will discuss this next.

Using the Marshal class for memory management

The `Marshal` class in .NET is a part of the `System.Runtime.InteropServices` namespace and provides a collection of methods for interacting with unmanaged code. It offers a suite of methods for allocating unmanaged memory, copying memory blocks, and converting between managed and unmanaged types, along with various other utilities for interacting with unmanaged code. This class is essential for scenarios wherein .NET code needs to interoperate with native code, such as calling functions in unmanaged libraries, working with **Component Object Model** (**COM**) objects, or manually managing memory.

There are instances when managed code must access unmanaged resources such as native libraries, COM objects, and Win32 APIs. For example, a C# application might need to call a C++ library to utilize a functionality unavailable in .NET. In such cases, the C# `Marshal` class becomes essential. The `Marshal` class in C# offers methods for managed code to interact with unmanaged resources. These methods facilitate marshaling data between managed and unmanaged code, allocating unmanaged memory, and invoking unmanaged functions.

Marshaling data involves converting information between managed and unmanaged code. When calling an unmanaged function, the arguments must be formatted so that the function can be interpreted. Similarly, when receiving data from an unmanaged function, it needs to be converted into a format that managed code can understand. The `Marshall` class offers methods that simplify marshaling data between managed and unmanaged code.

The first method of the `Marshall` class that we will explore is `StructureToPtr`. This method copies the contents of a managed structure to an unmanaged memory block. Let us consider a scenario where we must pass a structure from managed code to an unmanaged function. We'll define a simple structure in C# and use `Marshal.StructureToPtr` to convert it to a pointer that can be passed to an unmanaged function:

```csharp
using System;
using System.Runtime.InteropServices;
static class Program
{
    // Define a simple structure
    [StructLayout(LayoutKind.Sequential)]
    public struct MyStruct
    {
        public int x;
        public double y;
    }

    // Simulated unmanaged function
    [DllImport("NativeLibrary.dll",
    CallingConvention = CallingConvention.Cdecl)]
    public static extern void UnmanagedFunction(IntPtr ptr);

    static void Main()
    {
        MyStruct myStruct = new MyStruct
        {
            x = 42,
            y = 3.14
        };

        // Allocate unmanaged memory for the structure
        IntPtr ptr = Marshal.AllocHGlobal(Marshal.SizeOf(myStruct));

        try
        {
            // Convert the structure to a pointer
            Marshal.StructureToPtr(myStruct, ptr, false);
```

```
            // Call the unmanaged function with the pointer
            UnmanagedFunction(ptr);
        }
        finally
        {
            // Free the unmanaged memory
            Marshal.FreeHGlobal(ptr);
        }
    }
}
```

In the code snippet, the `MyStruct` structure is defined with two fields: an integer and a double. The `StructLayout` attribute ensures that the structure has a sequential layout in memory, which is necessary for interoperability with unmanaged code. We simulate creating an unmanaged function with `UnmanagedFunction`, which is declared using a `DllImport` attribute. This attribute imports unmanaged functions from a DLL or shared library. The attribute takes the name of the DLL or shared library as an argument and the function's name to import. `UnmanagedFunction` simulates an unmanaged function that takes a pointer to `MyStruct`. `Marshal.AllocHGlobal` allocates unmanaged memory that is sufficient to hold the structure. The size of the structure is determined using `Marshal.SizeOf`.

Conversely, we also have access to a `Marshal.PtrToStructure` method, which allows us to copy the contents of an unmanaged memory block to a managed structure. This can be used when retrieving data from an unmanaged function that returns a pointer to a structure. The following is an example of this method being used:

```csharp
[StructLayout(LayoutKind.Sequential)]
struct MyStruct
{
    public int a;
    public int b;
}

[DllImport("mylib.dll")]
static extern IntPtr UnmanagedFunction();

IntPtr ptr = UnmanagedFunction();
MyStruct myStruct = Marshal.PtrToStructure<MyStruct>(ptr);

Console.WriteLine("a = " + myStruct.a);
Console.WriteLine("b = " + myStruct.b);
Marshal.FreeHGlobal(ptr);
```

Here, we define a managed structure named MyStruct with two fields. Then, we have an unmanaged function named UnmanagedFunction, which returns a pointer to MyStruct. The UnmanagedFunction function is called and returns a pointer to an unmanaged memory block containing MyStruct. Marshal.PtrToStructure is then used to copy the contents of the unmanaged memory block to a managed instance of MyStruct. Finally, we print the values of the x and y fields of the MyStruct instance.

Notice that MyStruct is decorated with a StructLayout attribute. This attribute is used to control the physical layout of a class's data fields or a memory structure. It takes a LayoutKind enumeration (enum) to specify the layout type. There are three primary LayoutKind values:

- LayoutKind.Sequential ensures that the fields appear in memory in the same order as declared in the class or structure. It is commonly used when interacting with unmanaged code that expects a specific order of fields.

- LayoutKind.Explicit is used when you need to specify the exact memory offsets of fields, providing the most control over the layout. Each field must be annotated with the FieldOffset attribute to indicate its position in the memory layout.

- LayoutKind.Auto is the default layout kind and is typically used when you do not need to control the layout of the fields. The CLR will decide the most efficient layout for the fields. This layout kind is unsuitable for interop scenarios because the layout is not guaranteed to be consistent across different versions or implementations of the runtime.

In these examples, we saw the use of Marshal.AllocHGlobal and Marshal.FreeHGlobal, which are utility methods that facilitate allocating and freeing unmanaged memory. These methods are useful when working with unmanaged data that needs to be manipulated by managed code. Marshal.AllocHGlobal can allocate a block of unmanaged memory of a specified size and return a pointer to the allocated memory block:

```
IntPtr ptr = Marshal.AllocHGlobal(2048);
// use the allocated memory block
Marshal.FreeHGlobal(ptr);
```

Here, we allocate an unmanaged memory block of 2,048 bytes using Marshal.AllocHGlobal, use the allocated memory block for some operations, and then free the allocated memory block using the Marshal.FreeHGlobal method. The Marshal.FreeHGlobal method can be used to free a block of unmanaged memory that was allocated using the Marshal.AllocHGlobal method.

The Marshal class provides several powerful methods for working with unmanaged memory and calling unmanaged functions from managed code. These methods allow developers to allocate and deallocate memory with unmanaged code safely and efficiently. As mentioned several times before, always be cognizant of the potential pitfalls of this type of development. By following best practices and using the Marshal class correctly, robust and reliable applications can be built using unmanaged code.

Now that we have a fundamental understanding of how unmanaged memory can be allocated and deallocated, we will explore developing C# programs that depend on external and unmanaged code libraries.

Interoperability with unmanaged code

Interoperability in programming refers to the ability of two or more languages to interact as part of the same system. This generally means passing data between systems that have potentially been developed with different languages. This can include applications or libraries written in languages such as C or C++, which interact directly with the operating system's API. Since the integrated system does not run under the CLR's supervision, it is considered unmanaged code.

Interoperability is essential when you need to do any of the following:

- Utilize existing unmanaged libraries (e.g., DLLs written in C or C++)

- Access system-level functionalities not available in .NET

- Integrate with legacy code bases without rewriting them in managed languages

The theory supporting the need for interoperability is straightforward. Since we already understand unmanaged code, let's explore the first method of handling interoperability via P/Invoke.

P/Invoke

P/Invoke is a .NET feature that facilitates the calling of native functions implemented in unmanaged libraries from managed code. Examples include the Windows API or third-party libraries written in other languages. It allows .NET applications to specify the signatures of unmanaged functions that they wish to call through the `DllImport` attribute, which enables the .NET runtime to locate and invoke the corresponding function in the unmanaged DLL.

We saw similar code earlier in this chapter when we reviewed using the `Marshal` class. The following code shows a more practical example. We can allow a console application to access the Windows API and display a message box:

```
using System.Runtime.InteropServices;

// Declare the MessageBox function from user32.dll
[DllImport("user32.dll", CharSet = CharSet.Auto)]
static extern int MessageBox(IntPtr hWnd, string lpText, string
lpCaption, uint uType);

// Call the MessageBox function
MessageBox(IntPtr.Zero, "Hello, World!", "P/Invoke Example", 0);
```

We invoke the `MessageBox()` method found in the already-compiled `user32.dll`. *Figure 7.5* shows the resulting message box.

Figure 7.5 – The message box that is invoked by user32.dll

If the `DllImport` attribute was not used above the method, we would get the compile error displayed in *Figure 7.6*.

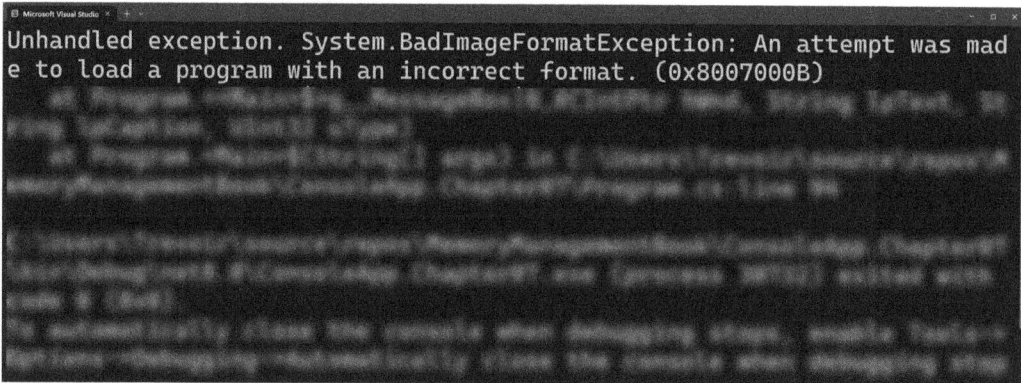

Figure 7.6 – The error that is shown when interop code is executed without DllImport

The `DllImport` attribute is used to specify the location of the unmanaged function. It includes several essential parameters:

- `DllName`: The name of the DLL containing the unmanaged function (e.g., `user32.dll`)
- `CharSet`: Specifies the character set that the unmanaged function uses (e.g., `CharSet.Auto` for automatic selection based on the platform)
- `CallingConvention`: Specifies the calling convention that the unmanaged function uses (e.g., `CallingConvention.Cdecl`)

Consider another example: you may need to interact with a SQLite database and its native code. SQLite is a popular, lightweight, disk-based database engine that integrates easily into applications. SQLite is written in C and exposes its functionality through a C API. To use SQLite in a .NET application without relying on third-party .NET wrappers (such as Entity Framework Core), you can use P/Invoke to call SQLite functions directly. This approach is helpful for scenarios wherein you need fine-grained control over database operations or want to minimize dependencies.

We will approach this implementation in three main stages:

1. Retrieve the SQLite C libraries. These can be downloaded from the official SQLite website (`https://www.sqlite.org/download.html`) and placed in a centrally accessible folder. During development, it is recommended that they be placed inside the project's folder.

2. Understand the SQLite C API. Identify the functions you need to call and their signatures. Here are some SQLite functions that we will call:

   - `sqlite3_open`: Opens a database connection

   - `sqlite3_close`: Closes a database connection

   - `sqlite3_exec`: Executes an SQL statement

3. Declare the SQLite functions in your .NET application using the `DllImport` attribute. This allows the .NET runtime to locate and invoke these functions from the SQLite library.

4. Call the functions. Use the declared functions in your .NET code to interact with the SQLite database.

In the code snippet that follows, we define the SQLite functions as stipulated in *step 2*:

```
using System.Runtime.InteropServices;
public static class SQLite
{
    [DllImport("sqlite3.dll", CallingConvention =
    CallingConvention.Cdecl)]
    public static extern int sqlite3_open(string filename,
    out IntPtr db);

    [DllImport("sqlite3.dll", CallingConvention =
    CallingConvention.Cdecl)]
    public static extern int sqlite3_close(IntPtr db);

    [DllImport("sqlite3.dll", CallingConvention =
    CallingConvention.Cdecl)]
    public static extern int sqlite3_exec(IntPtr db, string sql,
    IntPtr callback, IntPtr arg, out IntPtr errmsg);
}
```

Then, we continue to *step 3* and use these functions to interact with an SQLite database:

```
IntPtr db;
IntPtr errmsg;

// Open the database
if (SQLite.sqlite3_open("test.db", out db) != 0)
{
    Console.WriteLine("Failed to open database");
    return;
}

// Execute an SQL statement
string sql = "CREATE TABLE IF NOT EXISTS Test (Id INTEGER PRIMARY KEY,
Name TEXT)";
if (SQLite.sqlite3_exec(db, sql, IntPtr.Zero, IntPtr.Zero, out errmsg)
!= 0)
{
    Console.WriteLine("Failed to execute SQL: " +
    Marshal.PtrToStringAnsi(errmsg));
    SQLite.sqlite3_close(db);
    return;
}

Console.WriteLine("Table created successfully");

// Close the database
SQLite.sqlite3_close(db);
```

In this scenario, we omitted recommended error handling and memory management practices, so in implementing similar code, ensure that you do the following:

- Check the return codes of SQLite functions. Non-zero return codes indicate errors.

- When dealing with unmanaged code, be cautious about memory allocation and deallocation. For example, if `sqlite3_exec` returns an error message, you must free it using the appropriate SQLite function.

There may also be times when communication needs to be facilitated and directed in a P/Invoke call. For this, we need to implement a **callback function**. A callback function is a code in a managed application that assists an unmanaged DLL function in completing a task. Calls to a callback function are routed indirectly from the managed application, through the DLL function, and back to the managed implementation. Many DLL functions invoked via P/Invoke require a callback function in managed code to operate correctly.

To call most DLL functions from managed code, you must create a managed definition of the function and then call it. This process is generally straightforward, as seen here:

```
using System.Runtime.InteropServices;

// Define the callback delegate
delegate void CallbackDelegate(int value);

// Import the unmanaged function from the DLL
[DllImport("callback_example.dll", CallingConvention =
CallingConvention.Cdecl)]
private static extern void ProcessArray(int[] array, int size,
CallbackDelegate callback);

// Define the callback function
static void MyCallbackFunction(int value)
{
    Console.WriteLine("Processing value: " + value);
}

static void Main()
{
    int[] array = { 1, 2, 3, 4, 5 };

    // Create an instance of the callback delegate
    CallbackDelegate callbackDelegate =
    new CallbackDelegate(MyCallbackFunction);

    // Call the unmanaged function, passing the array and the callback
    // delegate
    ProcessArray(array, array.Length, callbackDelegate);
```

Always consult the DDL function's documentation to determine whether it requires a callback. Then create the callback function within your managed application.

Now that we know how to use P/Invoke for interop development, let's review another approach: the COM interop method.

Using COM interop

COM is another standard technology for interoperability. .NET provides robust support for interacting with COM components. You will typically start by adding a reference to the COM library, which generates an interop assembly. Some popularly used COM libraries are as follows:

- The Microsoft Office interop libraries enable .NET applications to automate and interact with Microsoft Office applications such as Excel, Word, and Outlook. Some libraries include `Microsoft.Office.Interop.Excel`, `Microsoft.Office.Interop.Word` and `Microsoft.Office.Interop.Outlook`.

- **ActiveX Data Objects (ADO)** is a set of APIs for database connectivity and operations, allowing interaction with different data sources such as SQL Server, Oracle, and Access. **ActiveX Data Objects Database (ADODB)** enables applications to perform operations such as querying, updating, and managing database connections.

- **Windows Script Host (WSH)** provides Windows scripting capabilities. It allows .NET applications to execute scripts and access various scripting engines. The `IWshRuntimeLibrary` allows .NET applications to interact with WSH, enabling automation of tasks such as creating shortcuts, manipulating the registry, and running scripts.

- DirectShow is a multimedia framework and API produced by Microsoft for video and audio streaming. It is used to perform various multimedia tasks on Windows. The `QuartzTypeLib` allows access to DirectShow features for tasks such as capturing video from a camera, rendering video files, and applying filters to media streams.

- Shell32 is a library that provides a way to interact with the Windows shell. It allows .NET applications to perform various tasks related to the Windows operating system's shell, such as managing files and folders, handling shell links, and manipulating the desktop environment. `Shell32.Shell` provides methods for tasks such as creating shortcuts, browsing for folders, and accessing special folders.

- The Windows Media Player library provides COM interfaces for interacting with the Windows Media Player application, allowing .NET applications to control media playback, manage playlists, and handle media events. `WMPLib.WindowsMediaPlayer` provides the functionality to embed and control Windows Media Player, manage media libraries, and handle playback controls.

- The Internet Explorer COM library allows .NET applications to control and automate the Internet Explorer browser. This is useful for web automation, testing, and scraping. `SHDocVw.InternetExplorer` provides methods and properties for managing the Internet Explorer browser, navigating to URLs, interacting with web pages, and handling browser events.

  Microsoft XML Core Services (MSXML) is a set of services that allow applications to build XML-based applications. It provides a way to parse, validate, and transform XML documents. `MSXML2.DOMDocument` provides functions to parse and manipulate XML documents, perform XSLT transformations, and execute XPath queries.

- The **Speech API (SAPI)** allows .NET applications to use speech recognition and text-to-speech capabilities through the `SpeechLib.SpVoice` library. `SpeechLib.SpInProcRecoContext` supports speech recognition capabilities, allowing applications to recognize spoken words and phrases.

To use a COM component in .NET, you will typically add a reference to the COM library, as shown in *Figure 7.7*. This generates an interop assembly to interact with Excel.

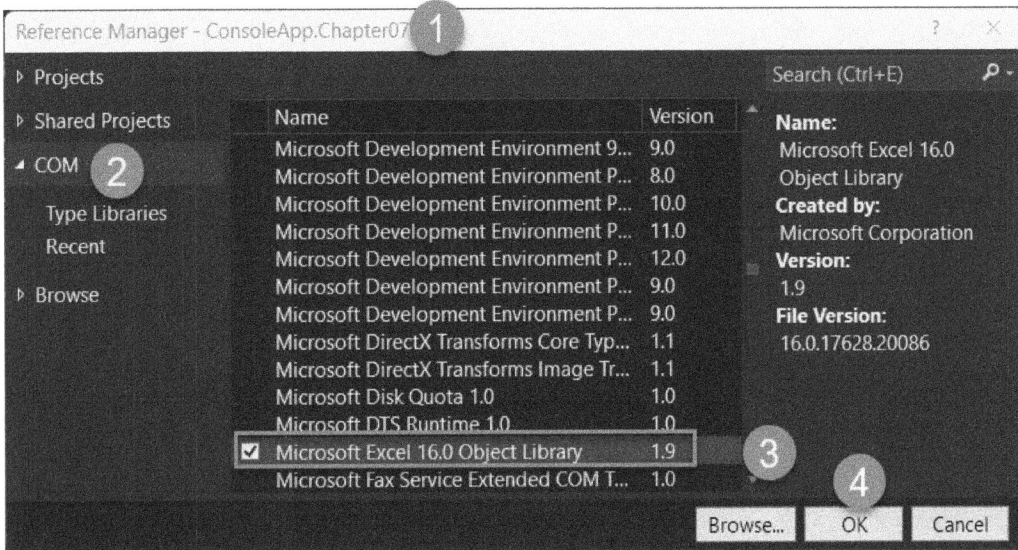

Figure 7.7 – Adding a COM reference to a C# project

If you are not using Visual Studio, then you can add the following to the `csproj` file, being mindful of the values that are placed in the `VersionMajor` and `VersionMinor` nodes:

```
<ItemGroup>
  <COMReference Include="Microsoft.Office.Interop.Excel">
    <WrapperTool>tlbimp</WrapperTool>
    <VersionMinor>9</VersionMinor>
    <VersionMajor>1</VersionMajor>
    <Guid>00020813-0000-0000-c000-000000000046</Guid>
    <Lcid>0</Lcid>
    <Isolated>false</Isolated>
    <EmbedInteropTypes>true</EmbedInteropTypes>
  </COMReference>
</ItemGroup>
using System;
using Excel = Microsoft.Office.Interop.Excel;
// Create a new instance of Excel application
```

```
Excel.Application excelApp = new Excel.Application();
excelApp.Visible = true;

// Add a new workbook
Excel.Workbook workbook = excelApp.Workbooks.Add();
Excel.Worksheet worksheet = workbook.Sheets[1];

// Write to the first cell
worksheet.Cells[1, 1].Value = "Hello, Excel!";

// Cleanup
workbook.Close(false);
excelApp.Quit();
System.Runtime.InteropServices.Marshal.ReleaseComObject(excelApp);
```

The `System` namespace provides fundamental and base classes that define commonly used data types, events, and handlers. `Excel = Microsoft.Office.Interop.Excel` creates an `Excel` alias for the `Microsoft.Office.Interop.Excel` namespace, which is part of the Interop Assembly generated from the Excel COM library. We then create a new instance of the Excel application and make it visible to the user with `excelApp.Visible = true;`. We then add a new workbook and some content to the first cell of the first worksheet.

This demonstrates the use of COM interop to automate Excel tasks from a .NET application. Using this basic example, you can add COM libraries, as well as interact with and use the generated interop assembly.

Another cross-cutting concern of interoperable code is that the data types between the languages, while similar in concept, may differ. For this, we must marshal the data types as the data is passed between them. We will discuss this next.

Interop marshaling

In software development, interoperability between different technologies and platforms is critical in ensuring the seamless functioning of complex systems. Interop marshaling in .NET is one such mechanism that allows managed code to interact with unmanaged code (such as native APIs written in C or C++). Interop marshaling is the process of transforming data types between managed and unmanaged environments. The .NET framework is a fundamental mechanism that facilitates communication between managed and unmanaged code. The .NET runtime, through the CLR, provides the necessary infrastructure to marshal data back and forth across the managed-unmanaged boundary.

In both managed and unmanaged memory, most data types share common representations, which are automatically managed by the interop marshaler. However, certain types are either ambiguous or lack representation in managed memory. Ambiguous types may have multiple unmanaged representations that correspond to a single managed type, or they might lack crucial information, such as the size of an array. The marshaler offers a default representation and alternative options where multiple representations are possible for these types. You can provide explicit instructions to guide the marshaler on handling an ambiguous type.

The CLR offers two primary mechanisms for interoperating with unmanaged code:

- **P/Invoke**: This allows managed code to call functions exported from an unmanaged library.

- **COM interop**: This enables managed code to interact with COM objects via interfaces.

Marshaling is necessary because the types of managed and unmanaged code differ. For example, in managed code, you might have a .NET string, which uses UTF-16 encoding, while unmanaged code might use various string encodings such as ANSI Code Page, UTF-8, null-terminated, or ASCII. By default, the P/Invoke subsystem handles these differences based on standard behavior. However, when you need more precise control, you can use the `MarshalAs` attribute to specify the expected type on the unmanaged side. For example, if you want to pass a string as a null-terminated UTF-8 string, you can do it as follows:

```
// Define the unmanaged function with the MarshalAs attribute
using System.Runtime.InteropServices;

[DllImport("example.dll", CallingConvention = CallingConvention.
Cdecl)]
static extern void UnmanagedFunction(
    [MarshalAs(UnmanagedType.LPStr)] string str);


string managedString = "Hello, World!";

// Call the unmanaged function
UnmanagedFunction(managedString);
```

The `[MarshalAs(UnmanagedType.LPStr)]` attribute specifies that the string should be marshaled as a null-terminated UTF-8 string (as it contains a reference to a **Long Pointer to String (LPStr)**). The `UnmanagedFunction` is called with the `managedString` argument, which will be marshaled according to the specified attribute. This ensures the string is correctly marshaled from the managed environment (UTF-16) to the unmanaged environment (UTF-8).

P/Invoke and COM interop use interop marshaling to accurately transfer method arguments between the caller, the callee, and back, if necessary. P/Invoke method calls flow from managed to unmanaged code and never the reverse, except in cases involving callback functions. Although P/Invoke calls are unidirectional (from managed to unmanaged code), data can flow in both directions as input or output parameters.

In contrast, COM interop method calls can flow in either direction. COM also employs a marshaler to transfer data between COM apartments or different COM processes. Only the interop marshaler is utilized when a call between managed and unmanaged code within the same COM apartment is made. However, when the call occurs between managed and unmanaged code in a different COM apartment or process, the interop marshaler and the COM marshaler are engaged.

As robust as the runtime's support for marshaling is, there are times when it isn't sufficient. In this situation, we use custom marshaling. Situations that require custom marshaling can include the following:

- When dealing with complex data structures that do not map directly to managed types (e.g., C# structs)

- When default marshaling introduces performance overheads that can be optimized with custom logic

- When data needs special handling during marshaling, such as custom transformations or validations

Custom marshaling is implemented by creating a class that implements the ICustomMarshaler interface. This interface defines methods for converting data between managed and unmanaged forms. What follows is an example of such an implementation:

```csharp
public struct ComplexData
{
    public int IntData;
    public string StringData;
}
public class ComplexDataMarshaler : ICustomMarshaler
{
    public object MarshalNativeToManaged(IntPtr pNativeData)
    {
        // Convert native data to managed ComplexData
        ComplexData data = new ComplexData();
        data.IntData = Marshal.ReadInt32(pNativeData);
        IntPtr stringPtr = Marshal.ReadIntPtr(pNativeData,
        IntPtr.Size);
        data.StringData = Marshal.PtrToStringAnsi(stringPtr);
        return data;
    }
```

```
    public IntPtr MarshalManagedToNative(object ManagedObj)
    {
        // Convert managed ComplexData to native data
        if (!(ManagedObj is ComplexData))
            throw new ArgumentException("ManagedObj is not of type
            ComplexData");

        ComplexData data = (ComplexData)ManagedObj;
        IntPtr pNativeData = Marshal.AllocHGlobal(IntPtr.Size * 2);
        Marshal.WriteInt32(pNativeData, data.IntData);
        IntPtr stringPtr = Marshal.StringToHGlobalAnsi
        (data.StringData);
        Marshal.WriteIntPtr(pNativeData, IntPtr.Size, stringPtr);
        return pNativeData;
    }

    public void CleanUpNativeData(IntPtr pNativeData)
    {
        // Clean up unmanaged string
        IntPtr stringPtr = Marshal.ReadIntPtr(pNativeData,
        IntPtr.Size);
        Marshal.FreeHGlobal(stringPtr);
        Marshal.FreeHGlobal(pNativeData);
    }

    public int GetNativeDataSize()
    {
        return IntPtr.Size * 2;
    }

    public static ICustomMarshaler GetInstance(string cookie)
    {
        return new ComplexDataMarshaler();
    }

    public void CleanUpManagedData(object ManagedObj)
    {
        throw new NotImplementedException();
    }
}

// Using the custom marshaler
```

```
[DllImport("SomeUnmanagedLibrary.dll")]
public static extern void ProcessComplexData([MarshalAs(UnmanagedType.
CustomMarshaler, MarshalTypeRef = typeof(ComplexDataMarshaler))]
ComplexData data);


ComplexData data = new ComplexData { IntData = 42, StringData =
"Hello, World!" };
ProcessComplexData(data);
```

The ICustomMarshaler interface requires the implementation of five methods:

- CleanUpManagedData(Object): This performs a cleanup of the managed data when it is no longer needed.

- CleanUpNativeData(IntPtr): This performs a cleanup of the unmanaged data when it is no longer needed.

- GetNativeDataSize(): This returns the size of the data to be marshaled. This is useful for memory allocation purposes.

- MarshalManagedToNative(Object): This converts the data from managed to unmanaged.

- MarshalNativeToManaged(IntPtr): This converts the data from unmanaged to managed

The preceding example depicts how a C# struct can be marshaled using custom marshaling. The struct has two fields based on primitive data types that could have been candidates for runtime-supported marshaling. Since they are wrapped by a more complex struct, we need to write custom logic to marshal each value before it can be used in the unmanaged code. Similarly, when the data is retrieved from the unmanaged code, we need custom logic to parse it back to the C# struct.

In keeping with the need for granular control over the marshaling operation, we use the MarshalAs attribute and specify MarshalAs(UnmanagedType.CustomMarshaler, MarshalTypeRef = typeof(ComplexDataMarshaler)) to indicate that we are using the custom marshaler to handle the operation.

Now that we have a better understanding of how to interact with unmanaged libraries and taxi data back and forth, we need to ensure that we have a safe way to dispose of the resources. We will review the SafeHandle class, another built-in mechanism for handling unmanaged resources.

Handling unmanaged resources with the SafeHandle class

Going back to earlier concepts, we can think of unmanaged resources as resources that are provisioned outside the GC's watchful gaze. This can include recounting earlier sections of the book: file handles, network connections, and memory allocated outside the .NET runtime. They pose the ever-present risk of resource leaks and application instability.

The `SafeHandle` class provides a reliable and secure way to handle unmanaged resources. It is an abstract class in the `System.Runtime.InteropServices` namespace that provides ways to ensure that resources are correctly released, even in the presence of exceptions or improper use. It allows us to ensure the following:

- **Automatic resource release**: Ensures that the resource is released when the handle is no longer needed, preventing resource leaks

- **Reliability**: Works correctly with the GC, ensuring that resources are released safely

- **Security**: Minimizes security risks by preventing handle recycling and invalid handle usage

- **Ease of use**: Provides a simple way to manage the life cycle of unmanaged resources, reducing the need for complex error-handling code

The `SafeHandle` class provides methods and properties to interact with unmanaged resources. We will attempt to create a file using the `kernel32.dll` library in the code snippet that follows. Notice how we can verify the validity of the `fileHandle` and wrap the native file stream resource in a scoped operation:

```
// Import the CreateFile function from kernel32.dll
[DllImport("kernel32.dll", CharSet = CharSet.Auto, SetLastError =
true)]
static extern SafeFileHandle CreateFile(
        string lpFileName,
        uint dwDesiredAccess,
        uint dwShareMode,
        IntPtr lpSecurityAttributes,
        uint dwCreationDisposition,
        uint dwFlagsAndAttributes,
        IntPtr hTemplateFile);

const uint GENERIC_READ = 0x80000000;
const uint OPEN_EXISTING = 3;

// Create a file handle
SafeFileHandle fileHandle = CreateFile(
    "example.txt",
    GENERIC_READ,
    0,
    IntPtr.Zero,
    OPEN_EXISTING,
    0,
    IntPtr.Zero);

if (!fileHandle.IsInvalid)
```

```
{
    using (FileStream fs = new FileStream(fileHandle,
    FileAccess.Read))
    {
        using (StreamReader reader = new StreamReader(fs))
        {
            string content = reader.ReadToEnd();
            Console.WriteLine(content);
        }
    }
}
else
{
    Console.WriteLine("Failed to open file.");
}
```

The `CreateFile` function from `kernel32.dll` is declared using P/Invoke. It returns a `SafeFileHandle` instead of an `IntPtr`. The `SafeFileHandle` is used to create a `FileStream`, which is then used to read the file contents. When the `SafeFileHandle` and `FileStream` objects are disposed of, the unmanaged file handle is released automatically, ensuring that there are no resource leaks.

Since the `SafeHandle` class is abstract, extending its functionality for custom operations through standard inheritance is possible. When you inherit from `SafeHandle`, you must override the `IsInvalid` and `ReleaseHandle()` members:

```
[DllImport("SomeUnmanagedLibrary.dll")]
static extern SafeExampleHandle GetExampleHandle();

SafeExampleHandle handle = GetExampleHandle();
if (!handle.IsInvalid)
{
    // Use the handle
    Console.WriteLine("Handle acquired and valid.");

    // The handle will be released when it goes out of scope and is
    // disposed
}
else
{
    Console.WriteLine("Failed to acquire handle.");
}

public class SafeExampleHandle : SafeHandle
```

```
{
    // Constructor
    public SafeExampleHandle() : base(IntPtr.Zero, true) { }

    // Override IsInvalid
    public override bool IsInvalid
    {
        get { return this.handle == IntPtr.Zero; }
    }

    // Override ReleaseHandle
    protected override bool ReleaseHandle()
    {
        // Perform necessary cleanup of the handle
        // For example, if the handle was allocated using Marshal.
        AllocHGlobal:
          Marshal.FreeHGlobal(this.handle);
          return true;
    }
}
```

A public parameter-less constructor must always be present. It should call the base constructor with a value representing an invalid handle value and a boolean value indicating whether the SafeHandle owns the native handle or not. In this case, the constructor initializes the base with IntPtr.Zero and a flag indicating that the handle is owned by the SafeHandle instance. We also override the IsInvalid property with custom logic to determine whether the handle is invalid. The ReleaseHandle method override performs the necessary cleanup, which, in this case, consists of freeing the memory allocated with Marshal.AllocHGlobal.

Now that we have a fundamental understanding of integrating with and handling unmanaged code and libraries, let us review the major talking points of this chapter.

Summary

This chapter provided an in-depth look at various techniques and concepts for low-level programming, primarily involving authoring unsafe code and code that interacts with unmanaged code. It covered essential topics such as memory allocation, pointers, fixed statements, and interop mechanisms, including P/Invoke, COM Interop, Interop Marshaling, and the SafeHandle class.

We started by reviewing what managed and unmanaged memory are and how we can allocate memory that the GC does not manage. This becomes essential when we have operations that need to be as performant as possible, which means we want to reduce the performance overhead of having the GC involved in allocating and deallocating the memory. This consequently also means that we must be more vigilant when writing this code as we introduce the risk of memory leaks, among other risks. Mechanisms such as `stackalloc` and the `Marshal` class give us functions that support the allocation and clean-up operations. Still, the developer must be deliberate and clearly understand the required functionality before considering these allocation methods, as we learned in this chapter.

We then reviewed what unsafe code is and how the default mode of the .NET runtime restricts unsafe code from being executed. Code is considered unsafe when it is implemented in a block marked with the `unsafe` keyword. This means that we are explicitly excluding the code from the supervision of the GC and attempting to use unmanaged data types such as pointers. Pointers help us directly access and manipulate memory spaces and their data, which is not a default offering. Once again, this can become necessary and useful when performance is critical in the application being developed. Still, as we learned in this chapter, it introduces several risks and must be used with extreme caution and knowledge.

Finally, we reviewed the interoperability features of C# and some possibilities. Interop is useful when we need to leverage functionality from an already compiled library that contains functionality that we may not want to attempt to recreate in our classes. This is another example of interacting with unmanaged code, where the GC cannot manage the code in the integrated library. We reviewed different methods for this kind of integration and discussed best practices.

To conclude, low-level programming involves writing code that interacts closely with the hardware and system resources, providing fine-grained control over memory and processor operations. This type of programming typically deals with direct memory manipulation, pointer arithmetic, and system-level operations. This technique requires a deep understanding of the computer's architecture and the operating system. Absolute care must be taken when employing these programming methods.

In the next chapter, we will review some general and architectural best practices for developing applications in specific environments.

8

Performance Considerations and Best Practices

Performance is critical in software development, influencing user satisfaction, resource utilization, and scalability. In the context of .NET development, memory management is crucial in achieving optimal performance. While the .NET runtime provides automatic memory management through its **garbage collector** (**GC**), developers must still be mindful of how their code allocates and uses memory to avoid performance bottlenecks.

Memory allocation and deallocation can be expensive operations, particularly in high-performance applications. Each allocation adds pressure to the GC, potentially leading to more frequent collections and increased application pauses. Understanding the cost of memory allocation and employing strategies to minimize unnecessary allocations are critical for maintaining performance.

This chapter aims to provide a comprehensive overview of performance considerations and best practices for memory management in .NET development. We will delve into the mechanisms of the .NET garbage collector, explore strategies for minimizing memory usage, and highlight common pitfalls that can lead to inefficient memory utilization.

We will explore the following tangible and applicable topics:

- Memory management in desktop environments
- Memory management in web environments
- Memory management in cloud environments

By the end of this chapter, you will have enough information to make the best decisions for the type of application you are engineering.

Technical requirements

- Visual Studio 2022 (`https://visualstudio.microsoft.com/vs/community/`)

- Visual Studio Code (`https://code.visualstudio.com/`)

- .NET 8 SDK (`https://dotnet.microsoft.com/en-us/download/visual-studio-sdks`)

Memory management in desktop environments

Memory management is crucial in any .NET application, but desktop environments present unique challenges and opportunities. Unlike web or mobile applications, desktop applications typically have a long lifecycle, more substantial user interactions, and often require handling large data sets or performing intensive computations. Some more specific nuances are as follows:

- **Long lifecycles**: Desktop applications often run for extended periods, sometimes days or weeks. This longevity increases the importance of effective memory management to prevent memory leaks and ensure sustained performance.

- **Rich user interactions**: Desktop applications usually offer more complex and rich user interfaces, leading to higher memory usage due to the extensive use of UI elements, images, and other resources.

- **Resource-intensive operations**: Many desktop applications perform heavy computations, handle large files, or manage significant amounts of data in memory, which can lead to substantial memory consumption.

- **User expectation for responsiveness**: Users expect desktop applications to be highly responsive and perform well, even under heavy load. Efficient memory management is critical to meeting these expectations.

- **Multithreading**: Desktop applications often use multiple threads to perform background tasks, improve responsiveness, and utilize multicore processors effectively. Managing memory in a multithreaded environment introduces additional complexity due to the need for thread-safe data structures and synchronization mechanisms.

- **Background processes**: Many desktop applications run background processes, such as auto-saving documents, synchronizing data, or monitoring system status. These processes must be managed carefully to avoid memory bloat and ensure they do not interfere with the main application's performance.

- **Resource sharing**: Desktop applications often share resources with other running applications, requiring them to be good citizens in the operating system environment. Efficient memory usage helps prevent the application from consuming excessive system resources and ensures that other applications run smoothly.

- **System APIs and libraries**: Desktop applications frequently interact with system APIs and external libraries, which may introduce additional memory management considerations, such as handling unmanaged resources, interfacing with COM objects, or using platform-specific features.

Now that we have a general idea of some of the nuances we need to be mindful of when developing desktop applications, let us review the first .NET framework for creating desktop applications: Windows Forms.

Using Windows Forms

Windows Forms is the oldest framework of the lot since it was a part of .NET Framework in its early days. It is a mature and straightforward UI framework for building Windows desktop applications and uses the traditional Windows API under the hood. Regarding resource management, WinForms applications tend to have higher memory usage due to the heavy reliance on Windows APIs and GDI/GDI+ for rendering. Each control in WinForms is a wrapper around a Windows handle (HWND), leading to increased memory consumption, especially with complex UIs. It also lacks built-in support for UI virtualization, meaning all UI elements are loaded into memory, even if they are not visible. This can lead to significant memory overhead in applications with large data sets or complex interfaces.

Creating a Windows Forms application is straightforward using Visual Studio 2022. Simply choose the Windows Forms App template from the project template selection, as seen in *Figure 8.1*. Alternatively, you can use the following command in an appropriate directory:

```
dotnet new winforms
```

Windows Forms development in Visual Studio features a **What You See is What You Get (WYSIWYG)** drag-and-drop editor. You must manually write the UI elements code if you use another editor (such as Visual Studio Code).

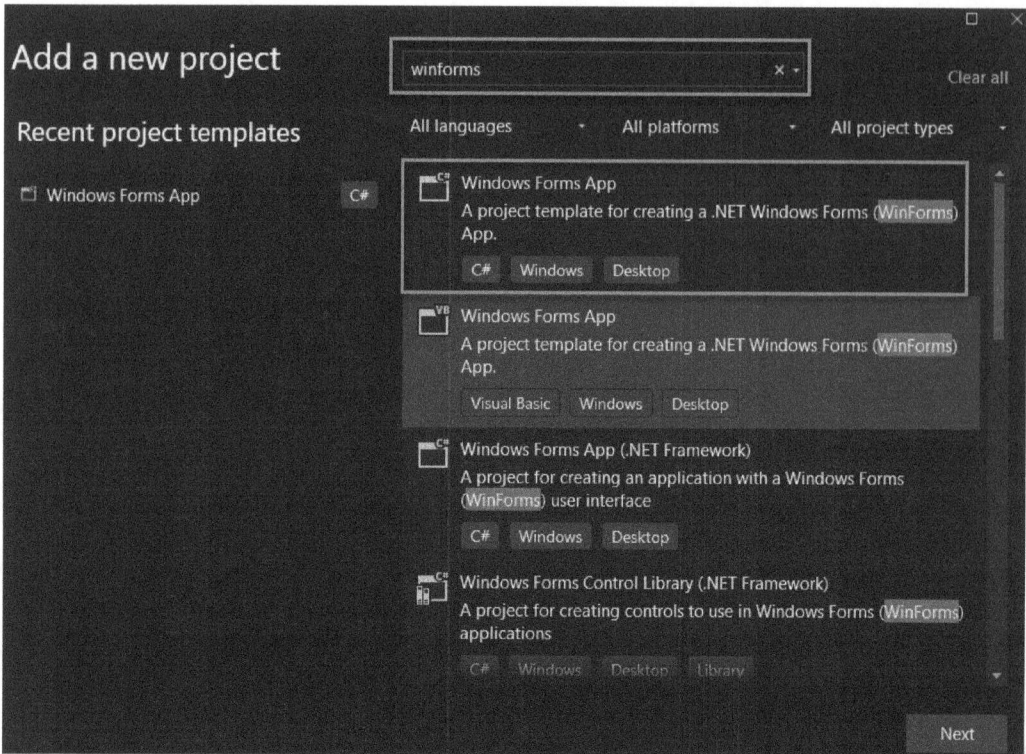

Figure 8.1 – Selecting the Windows Forms App project template

WinForms relies on the .NET GC for memory management. While it handles managed resources well, unmanaged resources (e.g., graphics objects) must be manually managed using the IDisposable pattern.

Notwithstanding its inefficiencies, WinForms is known for its simplicity and rapid development capabilities, making it a popular choice for simple or small-scale applications.

A tangible example of the need for the IDisposable pattern can be found in the scenario where we need to connect to a database. The database connection is an unmanaged resource, and releasing it when it's no longer needed is important. Implementing the IDisposable pattern ensures that the database connection is closed correctly.

Let us say that we have a class that encapsulates the database connection:

```
using Microsoft.Data.SqlClient;
namespace WinFormsApp.Chapter08;

public class DatabaseManager : IDisposable
{
    private SqlConnection _connection;
```

```csharp
private bool _disposed = false;

public DatabaseManager(string connectionString)
{
    _connection = new SqlConnection(connectionString);
    _connection.Open();
}

public SqlDataReader ExecuteQuery(string query)
{
    using (SqlCommand command = new SqlCommand(query,
    _connection))
    {
        return command.ExecuteReader();
    }
}


// Implement the Dispose method to close the connection
public void Dispose()
{
    Dispose(true);
    GC.SuppressFinalize(this);
}

protected virtual void Dispose(bool disposing)
{
    if (!_disposed)
    {
        if (disposing)
        {
            // Dispose managed resources
            if (_connection != null)
            {
                _connection.Close();
                _connection = null;
            }
        }

        // Dispose unmanaged resources here if needed

        _disposed = true;
    }
}
```

```
~DatabaseManager()
{
    Dispose(false);
}
}
```

You will see that the preceding method is similar to the previously explored `IDisposable` code examples in earlier chapters. In the `DatabaseManager` class, the `IDisposable` interface is implemented to manage the lifecycle of the database connection. The `Dispose` method closes the connection and releases resources.

The created project contains a default `Form1` form, which you can rename to `MainForm` to work with the following code. In this `MainForm`, drag a **DataGridsView** control onto the form as seen in *Figure 8.2*.

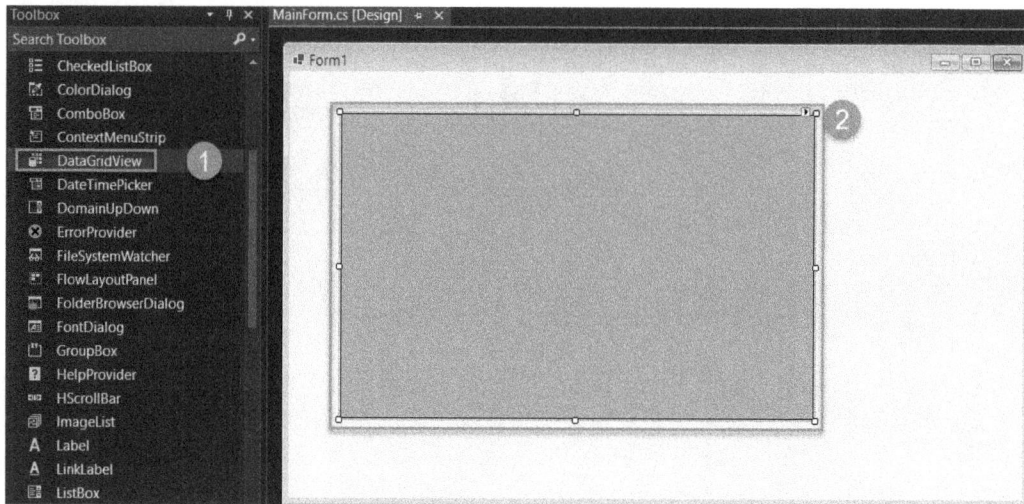

Figure 8.2 – Adding a DataGridView to a Windows form

In the `MainForm` code-behind, we can use the `DatabaseManager` class to manage the database connection and ensure proper disposal:

```
using Microsoft.Data.SqlClient;
using System.Data;

namespace WinFormsApp.Chapter08;

public partial class MainForm : Form
{
```

```
    private DatabaseManager _dbManager;

    public MainForm()
    {
        InitializeComponent();
        string connectionString = "your-connection-string-here";
        _dbManager = new DatabaseManager(connectionString);
    }

    private void btnLoadData_Click(object sender, EventArgs e)
    {
        string query = "SELECT * FROM YourTable";
        using (SqlDataReader reader = _dbManager.ExecuteQuery(query))
        {
            DataTable dataTable = new DataTable();
            dataTable.Load(reader);
            dataGridView1.DataSource = dataTable;
        }
    }

    // Ensure proper disposal when the form is closed
    protected override void OnFormClosed(FormClosedEventArgs e)
    {
        _dbManager.Dispose();
        base.OnFormClosed(e);
    }
}
```

In the MainForm class, the DatabaseManager instance is created and used to execute a query and load data into a DataGridView. The OnFormClosed method ensures that the DatabaseManager is disposed of when the form is closed, releasing the database connection properly.

Note that more secure and efficient ways exist to store the value for connectionString, and it is not recommended to write it directly in the code, as seen in the example. Additional steps for creating a database and constructing SQL statements for query have been omitted for brevity.

Now that we have reviewed some of the pros and cons of WinForms development, let us shift our focus to its successor, WPF.

Using Windows Presentation Foundation

Windows Presentation Foundation (WPF) was introduced with .NET Framework 3.0, a more modern UI framework designed to leverage the power of DirectX for rendering. It supports richer UI features, including advanced graphics, animations, and data binding. It uses a retained mode graphics system and leverages hardware acceleration through DirectX. This typically results in lower rendering memory usage than WinForms, especially for graphically intensive applications. WPF also provides built-in support for UI virtualization, particularly in controls such as `ListView`, `DataGrid`, and `TreeView`. This feature allows for more efficient memory usage by only loading visible elements into memory. The GC also plays a significant part in how WPF manages its resources, but its data binding and event handling can lead to memory leaks if not handled correctly. Proper use of `WeakEventManager` and understanding the WPF data binding lifecycle are crucial.

Let us start with creating a WPF application. This can be done by selecting the **WPF Application** project template in Visual Studio, as seen in *Figure 8.3*.

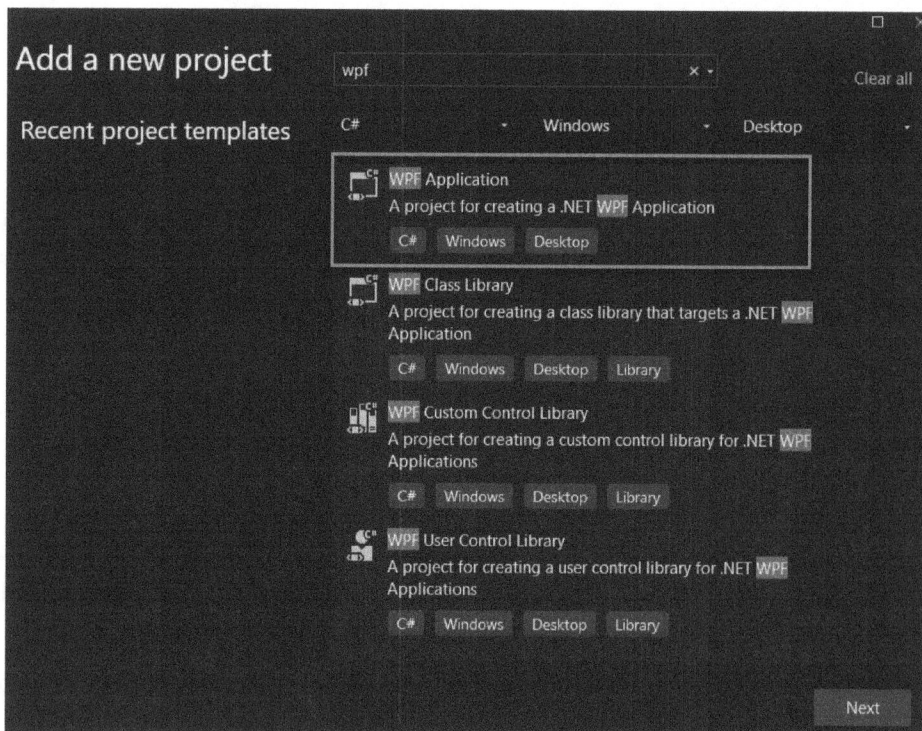

Figure 8.3 – Selecting the WPF Application template

To create this application using a CLI command, we use the following command in an appropriate folder:

```
dotnet new wpf
```

Like Windows forms development, Visual Studio offers a WYSIWYG editor for WPF. The UI can be generated using C#, but it primarily uses **Extensible Application Markup Language (XAML)**. It is like XML and is used to define rich controls for display and behavior purposes.

WPF event binding allows developers to handle events directly in XAML or code-behind. This is typically done to respond to user interactions, such as button clicks, text changes, and other UI events. In WPF, events are often routed events, meaning they can propagate up or down the visual tree, allowing for flexible and powerful event handling.

To bind an event in XAML, you typically use the event attribute of control and specify the event handler method defined in the code-behind. For example, we can add the following to the `MainWindow.xaml` file:

```xml
<Window x:Class="WpfApp.Chapter08.MainWindow"
        xmlns="http://schemas.microsoft.com/winfx/2006/xaml/
        presentation"
        xmlns:x="http://schemas.microsoft.com/winfx/2006/xaml"
        Title="MainWindow" Height="200" Width="300">
    <Grid>
        <Button Name="MyButton" Content="Click Me" Width="100"
        Height="30" VerticalAlignment="Center"
        HorizontalAlignment="Center" Click="MyButton_Click"/>
    </Grid>
</Window>
```

In `MainWindow.xaml.cs`, we use the following to register a click event on `MyButton`:

```csharp
private void MyButton_Click(object sender, RoutedEventArgs e)
{
    MessageBox.Show("Button clicked!");
}
```

While event binding is straightforward, improper handling can lead to memory leaks. Common pitfalls include the following:

- **Event handler not detached**: If an event handler is not detached from an event when an object is no longer needed, it can prevent the object from being garbage collected. This happens because the event subscriber holds a reference to the event handler, creating a strong reference chain that prevents garbage collection.

- **Anonymous delegates or lambda expressions**: Using anonymous delegates or lambda expressions can exacerbate the problem because these are harder to detach explicitly. They are often created inline and can be forgotten, leading to memory leaks.

- **Long-lived event publishers**: If the event publisher has a longer lifecycle than the subscriber (e.g., static events or events on singleton services), it can cause memory leaks if the subscriber is not properly unsubscribed.

WeakEventManager helps mitigate this problem by creating weak references, allowing objects to be collected by the GC when they are no longer needed.

To use the WeakEventManager, you typically need to do the following:

- Create a custom WeakEventManager for your event.

- Use this custom manager to subscribe and unsubscribe to events.

- Implement the IWeakEventListener interface in your event listener class.

Let us modify the code-behind or MainWindow.xaml.cs file to look like the following:

```
using System.Windows;
using System.Windows.Controls;

namespace WpfApp.Chapter08
{
    public partial class MainWindow : Window
    {
        public MainWindow()
        {
            InitializeComponent();
            WeakEventManager<Button, RoutedEventArgs>.
            AddHandler(MyButton, "Click", MyButton_Click);
        }

        private void MyButton_Click(object sender, RoutedEventArgs e)
        {
            MessageBox.Show("Button clicked!");
        }

        protected override void OnClosed(EventArgs e)
        {
            base.OnClosed(e);
            WeakEventManager<Button, RoutedEventArgs>.
            RemoveHandler(MyButton, "Click", MyButton_Click);
        }
    }
}
```

`WeakEventManager<Button, RoutedEventArgs>.AddHandler(MyButton, "Click", MyButton_Click);` subscribes to the Click event of `MyButton` using `WeakEventManager`. This ensures the event subscription does not create a strong reference to the `MainWindow` instance.

`WeakEventManager<Button, RoutedEventArgs>.RemoveHandler(MyButton, "Click", MyButton_Click);` in the `OnClosed` method unsubscribes from the event. It is a good practice to explicitly remove event handlers when the window is closed, ensuring no lingering references.

Using `WeakEventManager`, the `MainWindow` instance can be garbage collected even if the button's Click event is still being raised, as the subscription does not create a strong reference. Even though `WeakEventManager` helps with garbage collection, it's still good practice to unsubscribe explicitly when the window is closed.

WPF generally performs better when rendering complex UIs and handling large data sets. It also supports multi-threaded UI updates, further enhancing performance. This richer feature set has a steeper learning curve and a more complex development process than WinForms. An essential performance boost is the `WeakEventManager`, a powerful tool for managing event subscriptions in WPF applications to prevent memory leaks.

Another area for consideration is data binding when developing using the **Model-View-ViewModel (MVVM)** pattern. The MVVM pattern is commonly used in WPF, **Universal Windows Platform (UWP)**, and **Multi-platform App UI (MAUI)** applications to separate the presentation layer from the business logic, and while we will not go into the details of how it works and might be implemented, it is essential to know that it can introduce memory leaks and must be appropriately implemented. The general guidelines explored, such as using `WeakEventManager`, managing event handlers, proper use of async operations, and other techniques, help to mitigate these risks.

Now, we can move forward with memory management techniques for web environments.

Memory management in web environments

Developing web applications introduces specific memory management challenges distinct from other software development types. These challenges arise due to the inherent nature of web applications, which must handle multiple users, manage state across requests, and integrate with various services. Some of the unique memory management challenges faced when developing web applications and general solutions are as follows:

- **Concurrency and scalability**: Web applications often need to handle a high volume of concurrent requests from multiple users. Managing memory efficiently under such load is crucial to prevent the server from being overwhelmed. Poor memory management can lead to high memory consumption, increased garbage collection frequency, and ultimately, degraded application performance or crashes.

Implementing efficient request handling, connection pooling, and optimizing data structures for concurrent access can help manage memory better under high concurrency.

- **Session management**: Maintaining user state across multiple requests typically involves storing session data. This can consume significant memory, especially with many active sessions. Improper session management can lead to memory bloat, where memory usage steadily increases, potentially causing out-of-memory errors.

  Utilize session state efficiently, implement session timeouts, and consider using distributed caching solutions such as Redis or Memcached to manage session data outside of the web server's memory.

- **Resource allocation and cleanup**: Web applications frequently allocate and deallocate resources such as database connections, file handles, and network sockets. Properly managing these resources to ensure they are released when no longer needed is critical. Failure to release resources can result in resource leaks, leading to exhausted connections and degraded application performance.

  Implement proper resource cleanup using the `IDisposable` interface and the using statement in C#, and ensure all resources are correctly released after use.

- **Third-party integrations**: Integrating third-party services and libraries can introduce additional memory management complexities. Handling large data payloads and managing external resource lifecycles can be challenging. Memory leaks or excessive memory consumption can occur if third-party resources are not appropriately managed, affecting overall application performance.

  Use efficient data handling techniques, such as streaming large data sets rather than loading them entirely into memory, and ensure proper cleanup of third-party resources.

- **Asynchronous operations**: Web applications often perform asynchronous operations such as I/O tasks, background processing, and external API calls. Managing memory in asynchronous workflows requires careful attention to avoid leaks and dangling references. Improper handling of asynchronous operations can lead to increased memory usage and potential memory leaks, impacting application performance and stability.

  Use asynchronous programming patterns correctly, avoid capturing unnecessary references in asynchronous callbacks, and leverage `async/await` to manage asynchronous operations efficiently.

- **Distributed architectures**: Modern web applications frequently adopt microservices or distributed architectures, complicating memory management across multiple services and potentially across multiple servers. Inconsistent memory management practices across services can lead to fragmented memory usage, making tracking and optimizing overall memory consumption difficult.

Standardize memory management practices across services, use centralized logging and monitoring tools to track memory usage, and implement distributed caching and state management solutions.

- **Caching strategies**: Caching improves performance by reducing redundant computations and database queries. However, improper caching strategies can lead to excessive memory use. Over-caching or improper cache invalidation policies can cause memory bloat, where the cache grows uncontrollably, consuming significant memory.

  Implement efficient caching strategies, such as using cache expiration policies, sliding expiration, and cache eviction techniques to balance performance and memory usage.

- **Garbage collection tuning**: The GC plays a crucial role in memory management. Tuning the GC for optimal performance in a web environment can be challenging.

Understand and configure the GC settings appropriate for your application's workload, such as choosing between server and workstation GC modes and tuning generation thresholds for optimal performance.

Most of these topics have been addressed in previous chapters, and the same development techniques will be applied to manage memory and resource usage. Consider that each time an attempt is made to browse to a web location, resources are pooled and provisioned to support that request and response cycle. One of the most common operations carried out is usually database-related.

ADO.NET and **Entity Framework Core (EF Core)** are two primary data access technologies in the .NET ecosystem. Both are crucial in web development for different reasons, providing distinct advantages and trade-offs. Let us start by reviewing ADO.NET and some best practices for its usage.

Using ADO.NET

ADO.NET is a low-level data access technology that allows developers to interact directly with databases using SQL commands. It provides a set of classes for managing data, including `SqlConnection`, `SqlCommand`, `SqlDataReader`, and `SqlDataAdapter`.

The following components comprise the most prominent features:

- Connection: Manages the connection to the data source (e.g., `SqlConnection` for SQL Server).

- Command: Represents an SQL statement or stored procedure that can be executed against the database (e.g., `SqlCommand`).

- `DataReader`: Provides a way to read data from the data source in a forward-only, read-only manner (e.g., `SqlDataReader`).

- `DataAdapter`: Facilitates data retrieval and updating for disconnected data manipulation scenarios.

- `DataSet`: Represents an in-memory data cache suitable for disconnected data operations.

A typical ADO.NET operation involves opening a connection to a database, executing a query, and then fetching and processing a result for return. A typical example looks like the following code:

```
public void ADOExample()
{
    SqlConnection conn = new SqlConnection("your_connection_string");
    conn.Open();
    SqlCommand cmd = new SqlCommand("SELECT * FROM Users", conn);
    SqlDataReader reader = cmd.ExecuteReader();
    // Perform operations
    // Forgot to dispose resources
}
```

Based on the last comment in the preceding code block, you can see that the code is straightforward, but this writing style is not the best. This is where we discuss the disposal of resources after use. In this example, that conn is never disposed of and will remain open even after the request has been served. We can use the using statement to encapsulate the connection into a scope, which will automatically dispose of the object at the end of the operation:

```
public void ADOExample()
{
    using (SqlConnection conn = new SqlConnection
    ("your_connection_string"))
    {
        conn.Open();
        using (SqlCommand cmd = new SqlCommand("SELECT *
        FROM Users", conn))
        {
            using (SqlDataReader reader = cmd.ExecuteReader())
            {
                // Perform operations
            }
        }
    }
}
```

We have seen this pattern in previous chapters. Another important takeaway from the preceding code block is using SqlDataReader instead of DataTable. This can help with excessive memory usage. The alternative (and not recommended) way of writing this method in code would be as follows:

```
public void BadExample()
{
    using (SqlConnection conn = new SqlConnection
    ("your_connection_string"))
```

```
        {
            conn.Open();
            using (SqlCommand cmd = new SqlCommand("SELECT *
            FROM LargeTable", conn))
            {
                using (SqlDataAdapter adapter = new SqlDataAdapter(cmd))
                {
                    DataTable dataTable = new DataTable();
                    adapter.Fill(dataTable);
                    // Process data
                }
            }
        }
    }
```

`SqlDataReader` provides a fast, sequential, and read-only cursor for retrieving data from a database. It reads one row at a time and is designed for high-performance data access. Because `SqlDataReader` operates in a streaming mode, it does not load all data into memory simultaneously, resulting in a lower memory footprint. Given the cursor's sequential nature, it is well suited for processing large volumes of data where you do not need to go back and forth between records. `SqlDataReader` is generally faster than `DataTable` because it incurs less overhead regarding memory allocation and data management.

`DataTable`, on the other hand, loads all the data into memory, providing an in-memory representation of the entire result set. This makes it easier to manipulate and work with data disconnectedly. Unlike `SqlDataReader`, `DataTable` allows random access to any row or column in the result set, making it suitable for scenarios where data manipulation, updates, and complex operations such as sorting, filtering, and relationships between multiple tables are needed. Since `DataTable` stores all data in memory, it can consume excessive memory, especially with large data sets, affecting application performance and scalability.

ADO.NET is a robust data access technology that provides high performance and fine-grained control over database interactions. However, with this power comes the responsibility to manage resources effectively to avoid memory leaks and ensure optimal performance. By following these best practices—efficiently managing connections, disposing of unmanaged resources, optimizing data retrieval, using parameterized queries, and handling large data sets efficiently—developers can create web applications that are both performant and scalable. Understanding and applying these practices will help you make the most out of ADO.NET in your web development projects.

Now, let us review an alternative ORM called EF Core.

Using Entity Framework Core

Entity Framework Core (EF Core) is a modern, lightweight, open-source, and extensible version of the popular Entity Framework data access technology. EF Core is designed to work with .NET applications and provides an **Object-Relational Mapping (ORM)** framework, enabling developers to work with databases using .NET objects and LINQ queries. EF Core allows developers to interact with a database using a strongly typed object-oriented approach.

Since we already have a fundamental understanding of EF Core and LINQ from previous examples, we will focus on quick and easy optimizations for web applications. The first is using `AsNoTracking` for queries where the entities are not modified. This reduces the overhead of change tracking and improves performance and memory usage:

```
public List<User> QueryExample()
{
    using (var context = new AppDbContext())
    {
        return context.Users.AsNoTracking().ToList();
    }
}
```

Notice the combination of scope for the database context or connection object and the use of `AsNoTracking` on the read-only query operation. Tracking refers to the mechanism by which EF Core keeps track of the changes made to entities after they are retrieved from the database. When an entity is tracked, EF Core monitors it for property changes. This allows EF Core to automatically detect changes and generate the appropriate SQL commands to update the database when `SaveChanges()` is called.

In web applications, we are expected to perform read-only operations where data is fetched from the database and displayed to the user without modification. `AsNoTracking` improves query performance and memory usage by eliminating the overhead of tracking changes. This is particularly beneficial for high-traffic web applications where read operations are frequent. It also helps to reduce contention and locking issues, improving the overall throughput and responsiveness of the web application.

Another good practice is to avoid **eagerly loading** large related data sets and to use **explicit loading** or separate queries if necessary. Consider the following code:

```
public void EagerLoadingExample()
{
    using (var context = new AppDbContext())
    {
        var users = context.Users.Include(u => u.Orders).ToList();
        // Eager loading large data sets can cause high memory usage
    }
}
```

The related `Orders` for each user are being loaded in a single query. If thousands of users have several orders, this can become a huge dataset and fill memory with unnecessary data that might not be needed upfront. Eager loading is a mechanism by which EF Core loads related entities and the main entity in a single query. This is done using the `Include` and `ThenInclude` methods. These methods allow you to specify the related data that should be retrieved as part of the main query, ensuring that all necessary data is loaded up front. This can be as convenient as it can be dangerous.

It can be helpful to reduce database roundtrips by loading all required data in a single query, thus reducing the overhead associated with multiple queries.

Explicit loading, on the other hand, is the process of manually loading related data for an entity that has already been retrieved from the database. This is done using the Entry method of the `DbContext`, along with the `Collection` or `Reference` methods to load the related entities.

Explicit loading can reduce memory usage by loading related data only when needed rather than preloading it upfront. This is particularly useful in scenarios where memory resources are constrained.

We could rewrite the previous code example to use explicit loading as follows:

```
public void ExplicitLoadingExample()
{
    using (var context = new AppDbContext())
    {
        var users = context.Users.ToList();
        foreach (var user in users)
        {
            context.Entry(user).Collection(u => u.Orders).Load();
        }
        // Explicitly load related data only when needed
    }
}
```

We can load a user's related order if necessary, using this syntax. This will lead to increased database trips but reduce the overall footprint of the data loaded into memory at any given time.

Choosing between ADO.NET and EF Core depends on the specific needs of your web application. ADO.NET offers greater control and performance, making it suitable for complex, performance-critical applications. EF Core, on the other hand, provides a more abstract and developer-friendly approach, facilitating rapid development and maintainability.

Understanding the strengths and trade-offs of each technology enables developers to select the best tool for their specific scenarios, ensuring efficient and effective data access in their web applications. Either way, both abstractions deal with a database, but how do we optimize read/write operations and minimize database trips? The most common approach to this is caching. We will discuss this method next.

Caching patterns in ASP.NET Core

Caching is a fundamental technique to optimize performance and manage memory efficiently in software solutions. By storing frequently accessed data in a cache, applications can significantly reduce the load on primary data stores and improve response times.

Caching involves storing copies of frequently accessed data in a fast storage medium (usually in-memory) to retrieve them when needed quickly. Caches can be implemented at various levels, including the following:

- **Client-side**: Data is stored locally on the client device.

- **Server-side**: Data stored in memory on the application server.

- **Distributed caches**: Data stored in a distributed, in-memory store shared across multiple application servers.

Caching also helps applications to manage memory more efficiently. In-memory caches such as *Redis* or *Memcached* provide fast access to data while keeping the working data set in memory, significantly quicker than disk-based storage.

By caching frequently accessed data, applications can reduce the number of requests made to the primary database. This can significantly reduce the latency of retrieving data from disk-based storage systems, leading to fewer database queries and less CPU, memory, and network bandwidth usage.

We will base the following examples on using the *Redis* cache. Redis is a cross-platform, open-source, in-memory storage system that can be used as a distributed, in-memory key-value database, cache, and message broker. It can be downloaded at `https://redis.io/downloads/` and set up as a Docker container.

In a .NET application, you can use the `StackExchange.Redis` library to connect to Redis and perform operations. We can add this using the following command:

```
dotnet add package StackExchange.Redis
```

Next, you can set up a `RedisConnection` class to handle the caching operations and connection. Here, we will define the `ConnectionMultiplexer`, which is the central arbiter of the connection to Redis inside the CLR; it is best to maintain a single instance of `ConnectionMultiplexer` throughout its runtime:

```
public class RedisConnection
{
    private static Lazy<ConnectionMultiplexer> lazyConnection =
    new Lazy<ConnectionMultiplexer>(() =>
    {
        // Replace with your actual Redis connection string
        return ConnectionMultiplexer.Connect
```

```
        ("your_redis_connection_string");
    });

    public static ConnectionMultiplexer Connection =>
    lazyConnection.Value;
}
```

The Multiplexer is initialized with a connection string of the form HOST_NAME:PORT_ NUMBER,password=PASSWORD, where:

- HOST_NAME is the hostname of your server (localhost by default)

- PORT_NUMBER is the port number Redis is listening on (6379 by default)

- PASSWORD is your Redis server's password (e.g., secret_password)

Now, we can create a service that uses the connection and creates data in the cache using a key or retrieve it via a key. It will encapsulate all caching operations (setting, getting, and removing cache entries) and can be reused across different application parts:

```
using Newtonsoft.Json;
using StackExchange.Redis;

namespace Caching.Api.Chapter08;

public class CacheService
{
    private readonly IDatabase _cache;

    public CacheService()
    {
        _cache = RedisConnection.Connection.GetDatabase();
    }

    public async Task SetCacheAsync<T>(string key, T value,
    TimeSpan expiration)
    {
        var jsonData = JsonConvert.SerializeObject(value);
        await _cache.StringSetAsync(key, jsonData, expiration);
    }

    public async Task<T> GetCacheAsync<T>(string key)
    {
        var jsonData = await _cache.StringGetAsync(key);
        if (jsonData.IsNullOrEmpty)
```

```
        {
            return default(T);
        }
        return JsonConvert.DeserializeObject<T>(jsonData);
    }

    public async Task RemoveCacheAsync(string key)
    {
        await _cache.KeyDeleteAsync(key);
    }
}
```

Now, we can use this cache service as an alternative data source to query the database with each request. To interact with the database, we use the IProductRepository interface, defined as follows:

```
public interface IProductRepository
{
    public Task<Product> GetProductAsync(string Id);
}
```

ProductCatalogService uses IProductRepository for database operations and CacheService to handle caching logic, delegating the actual data retrieval from the repository only when necessary:

```
using Caching.Api.Chapter08.Repositories;
namespace Caching.Api.Chapter08.Services;

public record Product(string Id, string Name);

public class ProductCatalogService
{
    private readonly CacheService _cacheService;
    private readonly IProductRepository _productRepository;

    public ProductCatalogService(CacheService cacheService,
    IProductRepository productRepository)
    {
        _cacheService = cacheService;
        _productRepository = productRepository;
    }

    public async Task<Product> GetProductAsync(string productId)
    {
        var cacheKey = $"product:{productId}";
```

```
            var cachedProduct = await _cacheService.
            GetCacheAsync<Product>(cacheKey);

            if (cachedProduct != null)
            {
                return cachedProduct;
            }

            var product = await _productRepository.
            GetProductAsync(productId);
            await _cacheService.SetCacheAsync(cacheKey, product,
            TimeSpan.FromMinutes(30));

            return product;
        }
    }
```

In the `Program.cs` file, we can use the following lines of code to register `CacheService` and `ProductCatalogService` for use in the application:

```
builder.Services.AddSingleton<CacheService>();
builder.Services.AddTransient<ProductCatalogService>();
```

The following is an API controller injecting the `ProductCatalogService` and using it as an end point to retrieve product information:

```
using Caching.Api.Chapter08.Services;
using Microsoft.AspNetCore.Mvc;

namespace Caching.Api.Chapter08.Controllers;

[ApiController]
[Route("[controller]")]
public class ProductController : ControllerBase
{
    private readonly ProductCatalogService _productCatalogService;

    public ProductController(ProductCatalogService
    productCatalogService)
    {
        _productCatalogService = productCatalogService;
    }

    [HttpGet("{productId}")]
```

```
public async Task<IActionResult> GetProduct(string productId)
{
    var product = await _productCatalogService.
    GetProductAsync(productId);
    if (product == null)
    {
        return NotFound();
    }
    return Ok(product);
}
}
```

By the end of this setup, each request for product information will be handled in the following steps:

1. The request comes to the endpoint.

2. The `GetProductAsync()` method is called in the `ProductCatalogService` class.

3. The method attempts to retrieve the product information based on the `productid` value, which doubles as a cache key.

4. If the product was retrieved in the last thirty minutes, it will be quickly retrieved in the cache.

5. Otherwise, the database will be queried for the data, and the retrieved record will be cached for the next thirty minutes.

Caching is vital in memory management and overall performance optimization in cloud solutions. By reducing database load, enhancing scalability, improving memory efficiency, and providing high availability, caching ensures that cloud applications can handle demanding workloads and deliver a superior user experience.

Now, we can review some best practices for memory management in a cloud environment.

Memory management in cloud environments

While traditional desktop and web environments have established memory management practices, cloud environments introduce unique complexities. Desktop applications operate on a single machine with fixed memory and CPU resources. So, memory management techniques focus on maximizing the use of these static resources. Web applications run on servers with predictable resource limits. While scaling is possible, this often involves adding more servers or upgrading existing ones with relatively static resource allocation.

Developing a cloud application usually means that you want to facilitate dynamic scaling and resource allocation. Cloud applications must handle both vertical (adding more resources to an existing instance) and horizontal (adding more instances) scaling. This requires sophisticated memory management to ensure efficiency and cost-effectiveness.

We have reviewed some general instrumentations and monitoring practices in a previous chapter, and we took a detailed look at what cloud environments offer and some best practices. Managing memory and costs in the cloud involves more than employing efficient development practices; it also requires some knowledge of architectural design and cloud resource provisioning and management.

Here, we will focus on a mixture of development and architectural methods that can help us manage resources and reduce costs in a cloud solution. We will begin by looking into building efficient multi-tenant applications.

Multi-tenancy and memory management

Multi-tenancy is a core architectural principle in cloud computing, where a single instance of a software application serves multiple customers, known as tenants. Each tenant shares the same physical resources, but their data and configuration are isolated to ensure security and privacy. This approach offers significant advantages in terms of resource utilization, cost efficiency, and ease of maintenance.

Some tangible benefits of developing a multi-tenant application are as follows:

- **Maximized utilization of resources**: By sharing the same hardware, software, and networking infrastructure across multiple tenants. This leads to reduced costs since the economies of scale are achieved as resources are efficiently utilized, lowering the cost per tenant. Resources are also dynamically allocated based on demand, ensuring optimal performance without over-provisioning.

- **Simplified maintenance and upgrades**: With a single instance serving multiple tenants, maintenance and upgrades are simplified. Updates, patches, and upgrades are applied once to the shared infrastructure, reducing the complexity and cost of maintenance. All tenants operate on the same software version, ensuring consistency and simplifying support.

- **Scalability**: Multi-tenancy allows applications to scale seamlessly since additional tenants can be accommodated by adding more instances or resources without significant changes to the application architecture. Resources can be scaled up or down based on demand, providing flexibility for varying workloads.

A typical application is a single-tenant application. Single-tenant development refers to an architectural approach where each customer (or tenant) has a dedicated instance of an application and its associated resources. This model is straightforward compared to multi-tenancy and is often used for applications requiring high customization, security, or isolation levels. Each tenant has a separate instance of the application, including a dedicated database, storage, and computing resources. This generally leads to higher costs and scaling and security considerations.

A **Customer Relationship Management (CRM)** system is an excellent example of an application that can be developed using a multi-tenancy approach. This system can be developed as a SaaS platform to be used by various companies to manage their customer interactions, sales, and marketing activities.

Each client or tenant manages its data within the shared CRM system. Tenants can customize their workflows and processes using the same underlying infrastructure, which runs on shared servers and databases to optimize resource utilization. This is depicted in *Figure 8.4*:

```
                    +-------------------------+
                    |         Clients         |
                    |-------------------------|
                    | - Tenant 1              |
                    | - Tenant 2              |
                    | - Tenant 3              |
                    +-----------+-------------+
                                |
                                v
                    +-------------------------+
                    |      Load Balancer      |
                    +-----------+-------------+
                                |
        +-----------------------+-----------------------------+
        |                       |                             |
        v                       v                             v
+-------------------+ +-------------------+ +-------------------+
| Application Server| | Application Server| | Application Server|
|-------------------| |-------------------| |-------------------|
| - Handles requests| | - Handles requests| | - Handles requests|
| - Business logic  | | - Business logic  | | - Business logic  |
+---------+---------+ +---------+---------+ +---------+---------+
          |                     |                     |
    +-----------------+-----------------------+-------------+
          |                     |
          v                     v
+-------------------+ +-------------------+
|   Database Server | |   Shared Services |
+-------------------+ +-------------------+
| - Tenant 1 Schema | | - Authentication  |
| - Tenant 2 Schema | | - Logging         |
| - Tenant 3 Schema | | - Monitoring      |
+-------------------+ +-------------------+
```

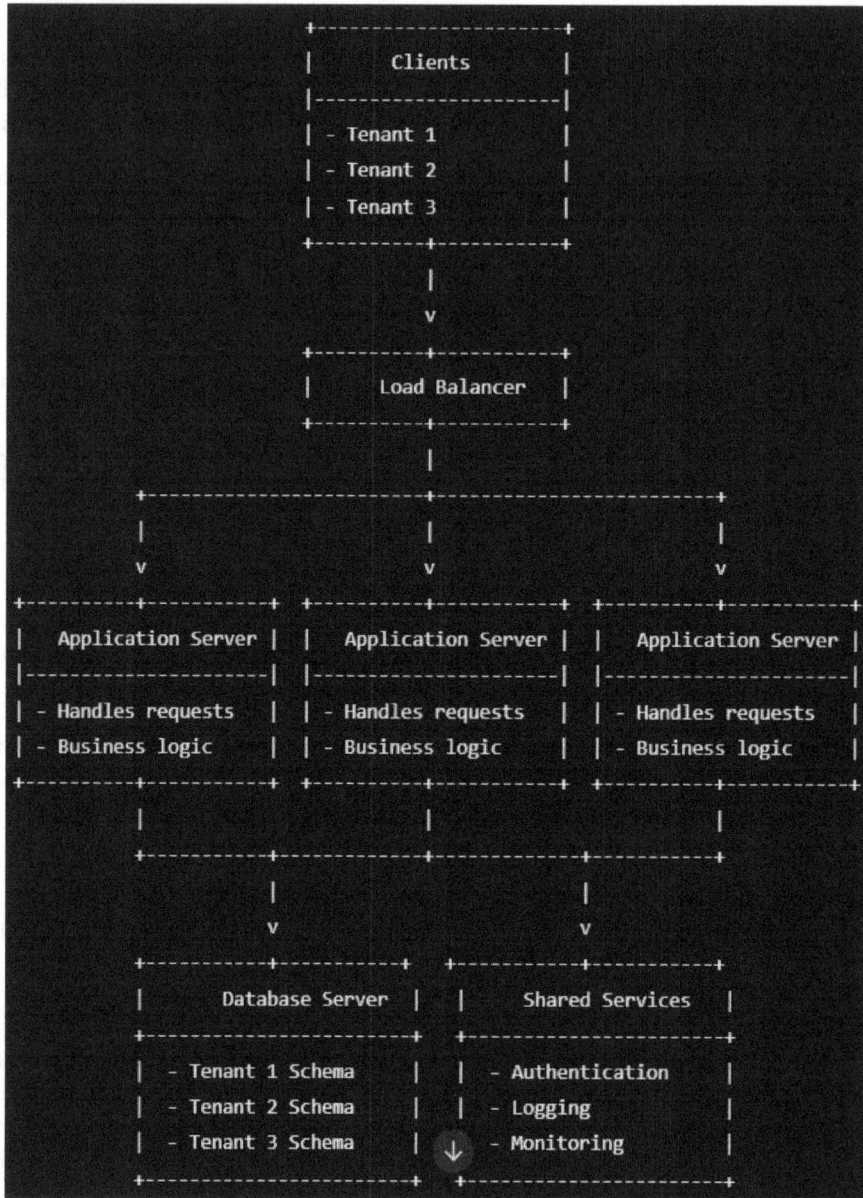

Figure 8.4 – Standard multi-tenant architecture

This architecture stores each tenant's data in separate schemas within the same database, ensuring data isolation and security; resources are dynamically allocated memory and CPU resources based on tenant usage patterns; and updates and maintenance tasks are performed centrally, ensuring all tenants benefit from improvements without downtime.

Since this application has a separate schema for each tenant, it is essential to dynamically select the appropriate schema based on the tenant making the request. First, let's set up a database context that dynamically switches schemas based on the tenant:

```
using Microsoft.EntityFrameworkCore;

namespace MultiTenant.Api.Chapter08.Data;

public class TenantDbContext : DbContext
{
    public readonly string TenantId;

    public TenantDbContext(DbContextOptions<TenantDbContext>
    options, string tenantId)
        : base(options)
    {
        TenantId = tenantId;
    }

    protected override void OnConfiguring(DbContextOptionsBuilder
    optionsBuilder)
    {
        var tenantSchema = $"tenant_{TenantId}";
        optionsBuilder.UseSqlServer($"Server=CrmServerAddress;
        Database=CrmDataBase;User Id=DbUsername;Password=DbPassword;
        SearchPath={tenantSchema}");
    }

    public DbSet<Customer> Customers { get; set; }
}
```

`TenantId` stores the tenant ID, which is used to identify the tenant and their specific schema. The constructor accepts standard database context options and a `tenantId`, which is then used to set the `TenantId` property.

`OnConfiguring` is overridden to configure the database context dynamically based on the tenant ID. Then, in the connection string, the schema name is defined using the tenant ID (`tenant_{TenantId}`).

The `Customer` class can be defined as follows:

```
public class Customer
{
    public int Id { get; set; }
    public string Name { get; set; }
}
```

We can inject the database context like usual and run queries as needed. The following is an example of using the context in an API endpoint:

```
namespace MultiTenant.Api.Chapter08.Controllers;

[Route("api/[controller]")]
[ApiController]
public class CustomersController : ControllerBase
{
    private readonly TenantDbContext _context;

    public CustomersController(TenantDbContext context)
    {
        _context = context;
    }

    // GET: api/customers
    [HttpGet]
    public async Task<ActionResult<IEnumerable<Customer>>>
    GetCustomers()
    {
        return await _context.Customers.ToListAsync();
    }
}
```

The database context is injected and used to access tenant-specific data. Now we need to use a middleware to dynamically retrieve the tenant ID from incoming requests, which can then be filtered down to the database context. This middleware looks as follows:

```
namespace MultiTenant.Api.Chapter08.Middleware;

public class TenantMiddleware
{
    private readonly RequestDelegate _next;

    public TenantMiddleware(RequestDelegate next)
    {
        _next = next;
    }

    public async Task InvokeAsync(HttpContext context,
    IServiceProvider serviceProvider)
    {
        var tenantId = context.Request.Headers["Tenant-ID"].
```

```
        ToString();

        if (string.IsNullOrEmpty(tenantId))
        {
            context.Response.StatusCode = 400; // Bad Request
            await context.Response.WriteAsync ("Tenant-ID header
            is missing");
            return;
        }

        var dbContextOptions = serviceProvider.GetRequiredService
        <DbContextOptions<TenantDbContext>>();
        context.RequestServices = serviceProvider.CreateScope().
        ServiceProvider;
        var tenantContext = new TenantDbContext(dbContextOptions,
        tenantId);
        context.RequestServices.GetRequiredService<IServiceScope>().
        ServiceProvider.GetService<TenantDbContext>();

        await _next(context);
    }
}
```

The `TenantMiddleware` class is a custom middleware component responsible for identifying the tenant based on the request headers and configuring the services accordingly. `InvokeAsync` is the core method of the middleware, and it handles the incoming HTTP request, determines the tenant, and configures the database context accordingly. For this to work, a request header called `Tenant-ID` must be included in the incoming HTTP request. After validating that the value is present, an instance of `TenantDbContext` is created by passing the `DbContextOptions` and `tenantId` parameters.

Ensure that the middleware is configured in the `Program.cs` file using the following line:

```
using MultiTenant.Api.Chapter08.Middleware;

var builder = WebApplication.CreateBuilder(args);
// Add services to the container.
builder.Services.AddControllersWithViews();

var app = builder.Build();
// Code omitted for brevity
app.UseMiddleware<TenantMiddleware>();
// Code omitted for brevity
app.Run();
```

Multi-tenancy is a great way to manage cloud resources and memory allocation in an architecture that is designed to handle high system load and is expected to boost maximum uptime. Now, we can look at another common approach to managing resources in a cloud and distributed system. This is called distributed caching.

Caching for memory management in cloud solutions

We previously looked at the fundamentals of caching and how it helps to speed up read operations in web applications. We can now consider using that same coding pattern to speed up and optimize our cloud solutions, ultimately reducing costs.

Cloud resources critical to the application's operation, including databases and storage, are often billed based on usage, reducing the number of database queries and improving memory efficiency. Caching helps lower operational costs. A cache is an additional cloud resource, but it can lead to cost savings in the long run.

Most, if not all, cloud providers have a native caching service, and Redis cache is at the helm of the service implementation. The code we wrote in the previous *Caching patterns for ASP.NET Core* section will work, so we will focus on how caching services can be provisioned.

The marquee caching service in Microsoft Azure is the **Azure Cache for Redis**. It is a fully managed, in-memory cache that enables high-performance and scalable architectures. Use it to create cloud or hybrid deployments. You can provision this service using the Azure portal or Azure CLI/PowerShell commands using Microsoft Azure. For this example, we will use the Azure CLI. This needs to be installed. You can find further instructions at `https://learn.microsoft.com/en-us/cli/azure/install-azure-cli#install`.

Once installed, use the following command to provide a high-availability **Azure Cache for Redis** service instance:

```
az redis server-link create --name myRedisCache --resource-group
myResourceGroup --server-to-link /subscriptions/{subscription-id}/
resourceGroups/{resource-group}/providers/Microsoft.Cache/Redis/
{linked-cache-name} --replication-role Secondary
```

Similarly, if AWS is your preferred cloud provider, your option is **Amazon ElastiCache**. This fully managed Redis OSS and Memcached-compatible service delivers real-time, cost-optimized performance for modern applications. You can similarly, install the AWS CLI by following this guide: `https://docs.aws.amazon.com/cli/latest/userguide/getting-started-install.html`. The following command can then be used to provision an `ElastiCache` service instance:

```
aws elasticache modify-replication-group --replication-group-id my-
replication-group --apply-immediately --automatic-failover-enabled
```

Some tangible benefits of using a caching service for a cloud solution are as follows:

- **Reduced database load**: Caching reduces the number of read operations on the database, which is particularly beneficial for read-heavy databases. This reduction in database load translates to lower costs regarding database instances, storage, and maintenance.

- **Faster data access**: Data stored in a cache is typically in memory, allowing for much faster access than fetching data from disk-based storage. This reduction in data transfer speeds up response times and lowers network bandwidth costs, which can be significant in data-intensive applications.

- **Improved scalability:** Caching helps applications scale more effectively by reducing the workload on backend systems. As the number of users increases, the cache can handle more requests without a corresponding increase in database queries, thus maintaining performance levels.

- **Efficient memory utilization**: Modern caching solutions offer advanced memory management features such as eviction policies (e.g., Least Recently Used (LRU)), which ensure that the most frequently accessed data remains in memory while less frequently accessed data is purged. By caching the results of complex database queries, applications can avoid repeating expensive operations.

Caching is an indispensable tool in the arsenal of cloud architects and developers. Organizations can significantly improve memory management and cost efficiency by strategically implementing caching mechanisms. From reducing database load and accelerating data access to optimizing resource allocation and lowering infrastructure costs, the benefits of caching are manifold.

As cloud solutions continue to evolve, the role of caching will only grow in importance, making it a critical component of any robust, cost-effective, and high-performance cloud strategy.

Now that we have reviewed some tangible cloud application design approaches, let us wrap up this chapter.

Summary

Memory management is a critical aspect of software development that directly impacts application performance, reliability, and scalability. In the .NET ecosystem, developers must consider different memory management strategies based on the application type being developed.

Desktop applications run on a single machine with finite resources. Efficient memory usage is crucial to avoid exhausting system resources. Understanding the .NET GC's behavior relative to the framework (Windows Forms, WPF, or WinUI) is extremely important. Each framework has nuances, so different coding techniques can be employed to ensure the application operates efficiently.

Web applications must handle multiple concurrent users, necessitating efficient memory usage to ensure responsiveness and scalability. A web application's most critical performance point is how it accesses data. In .NET, ADO.NET and Entity Framework Core are the two main libraries that facilitate database operations. Each has its efficiencies, and choosing between them can lead to significant performance gains in certain situations.

Leveraging caching mechanisms to reduce database load and improve response times while managing memory usage efficiently. This helps with web applications, especially if hosted in a cloud environment. When developing cloud solutions, most web development considerations come into play, but we must also consider architectural approaches that maximize provisioned resources and reduce long-term costs. Cloud environments support dynamic scaling, both vertically and horizontally, which impacts memory management strategies.

Multi-tenancy is an architectural approach that ensures memory isolation and efficient usage in multi-tenant applications to prevent one tenant from impacting others while optimizing memory usage to reduce costs since the cloud resources will be shared by several clients or tenants.

Effective memory management is crucial for the performance and reliability of .NET applications across desktop, web, and cloud environments. By understanding the unique challenges and adopting best practices tailored to each environment, developers can optimize memory usage, enhance application performance, and ensure scalability. This chapter provides a foundational understanding and actionable insights for developers to implement robust memory management strategies in their .NET applications.

9

Final Thoughts

In this book, we have explored the essential principles and practices of managing memory in the .NET development, providing a comprehensive guide for developers seeking to optimize their applications. We began with an introduction to memory management fundamentals, laying the groundwork for understanding how memory is allocated and managed within the .NET runtime environment.

In *Chapter 1*, we began our journey into the world of .NET memory management by laying the groundwork with essential concepts and principles. The chapter started with an overview of the importance of effective memory management in software development, highlighting how it directly impacts application performance and reliability.

Moving forward, we explored memory partitioning and allocation, detailing how the .NET framework organizes memory and allocates resources for efficient application performance. This section delved into the different memory segments, such as the heap and stack, and explained how memory is dynamically allocated during application runtime.

A crucial aspect of memory management, object lifetimes, and garbage collection was thoroughly covered. We thoroughly reviewed how the .NET garbage collector works, identifies and reclaims unused memory, and the best practices for managing object lifetimes to ensure optimal memory usage and application performance.

We then addressed common issues such as memory leaks and resource management, offering strategies to identify, prevent, and mitigate memory leaks in .NET applications. This section emphasized the importance of proper resource management and the techniques for ensuring that resources are efficiently utilized and released when no longer needed.

The content was then raised to a more advanced level, where we discussed advanced memory management techniques, providing a deeper dive into topics such as custom memory allocators, memory pools, and handling large objects. These advanced techniques help developers fine-tune their applications for specific performance requirements.

Memory profiling and optimization are key themes, with practical guidance on using profiling tools to identify memory usage patterns and bottlenecks. How applications can be reviewed and optimized through analyzing memory profiles and applying targeted optimizations provides valuable insight into future development endeavors.

Low-level programming was the most advanced topic, where we reviewed how to work with unmanaged code and interoperate with lower-level memory management APIs, expanding their ability to handle memory more precisely and efficiently.

The final practical chapter reviewed specific frameworks, performance considerations, and best practices for cloud, desktop, and web development. These best practices help developers tailor their memory management strategies to different application environments' unique requirements and constraints.

Effective memory management is the cornerstone of robust and high-performing applications in .NET and C#. These practices are crucial for small and large projects, as they help maintain application stability, improve performance, and reduce the likelihood of memory-related issues.

Be sure to use value types appropriately. Value types (int, float, and structs) are stored on the stack, whereas reference types (classes) are stored on the heap. Understanding the differences and using value types when appropriate can reduce heap allocations and improve performance by minimizing heap fragmentation and improving memory access speed.

Leverage the garbage collector efficiently. The .NET garbage collector automatically reclaims memory allocated to objects no longer in use. Writing code that supports efficient garbage collection, such as avoiding unnecessary object allocations and implementing proper object disposal, enhances performance.

Dispose of unmanaged resources where applicable. This ensures that resources such as file handles and database connections are released promptly. This prevents resource leaks, which is especially critical in large applications with significant resource management requirements.

Use the right collection types for your needs. For example, use `List<T>` instead of `ArrayList` for type safety and performance, use `Dictionary<K,V>` for fast lookups, and use `HashSet<T>` for unique items. This optimizes memory usage and access speed, which is crucial for applications handling large datasets.

Profile and monitor memory usage regularly using tools such as Visual Studio Profiler, .NET Memory Profiler, or JetBrains dotMemory. Monitoring helps identify memory leaks and inefficient memory usage and ensures ongoing optimization and the early detection of memory issues, which are crucial for maintaining performance in large applications.

Regardless of the size of the project, following these general practices and guidelines ensures that the application runs smoothly with limited resources. It also enhances application scalability and performance by reducing the risk of critical memory-related issues in production.

Developers can create more efficient, reliable, and maintainable .NET applications by implementing these best practices. Understanding and applying these principles from the outset helps ensure that small and large projects achieve optimal performance and robustness.

Index

V

W

‹packt›

packtpub.com

Subscribe to our online digital library for full access to over 7,000 books and videos, as well as industry leading tools to help you plan your personal development and advance your career. For more information, please visit our website.

Why subscribe?

- Spend less time learning and more time coding with practical eBooks and Videos from over 4,000 industry professionals

- Improve your learning with Skill Plans built especially for you

- Get a free eBook or video every month

- Fully searchable for easy access to vital information

- Copy and paste, print, and bookmark content

Did you know that Packt offers eBook versions of every book published, with PDF and ePub files available? You can upgrade to the eBook version at packtpub.com and as a print book customer, you are entitled to a discount on the eBook copy. Get in touch with us at customercare@packtpub.com for more details.

At www.packtpub.com, you can also read a collection of free technical articles, sign up for a range of free newsletters, and receive exclusive discounts and offers on Packt books and eBooks.

Other Books You May Enjoy

If you enjoyed this book, you may be interested in these other books by Packt:

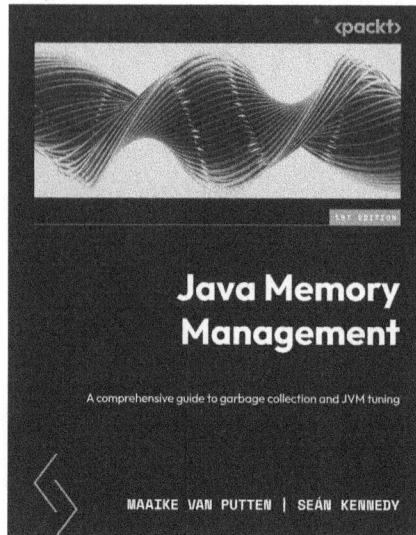

Java Memory Management

Maaike van Putten, Dr. Seán Kennedy

ISBN: 978-1-80181-285-6

- Understand the schematics of debugging and how to design the application to perform well
- Discover how garbage collectors work
- Distinguish between various garbage collector implementations
- Identify the metrics required for analyzing application performance
- Configure and monitor JVM memory management
- Identify and solve memory leaks

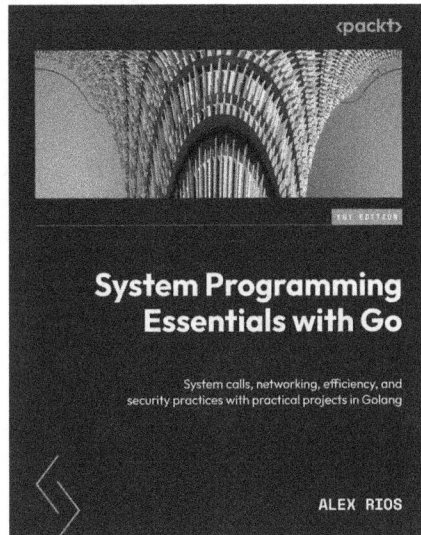

System Programming Essentials with Go

Alex Rios

ISBN: 978-1-83763-413-2

- Understand the fundamentals of system programming using Go
- Grasp the concepts of goroutines, channels, data races, and managing concurrency in Go
- Manage file operations and inter-process communication (IPC)
- Handle USB drives and Bluetooth devices and monitor peripheral events for hardware automation
- Familiarize yourself with the basics of network programming and its application in Go
- Implement logging, tracing, and other telemetry practices
- Construct distributed cache and approach distributed systems using Go

Packt is searching for authors like you

If you're interested in becoming an author for Packt, please visit `authors.packtpub.com` and apply today. We have worked with thousands of developers and tech professionals, just like you, to help them share their insight with the global tech community. You can make a general application, apply for a specific hot topic that we are recruiting an author for, or submit your own idea.

Share Your Thoughts

Now you've finished *Effective .NET Memory Management*, we'd love to hear your thoughts! Scan the QR code below to go straight to the Amazon review page for this book and share your feedback or leave a review on the site that you purchased it from.

`https://packt.link/r/1835461042`

Your review is important to us and the tech community and will help us make sure we're delivering excellent quality content.

Download a free PDF copy of this book

Thanks for purchasing this book!

Do you like to read on the go but are unable to carry your print books everywhere?

Is your eBook purchase not compatible with the device of your choice?

Don't worry, now with every Packt book you get a DRM-free PDF version of that book at no cost.

Read anywhere, any place, on any device. Search, copy, and paste code from your favorite technical books directly into your application.

The perks don't stop there, you can get exclusive access to discounts, newsletters, and great free content in your inbox daily

Follow these simple steps to get the benefits:

1. Scan the QR code or visit the link below

https://packt.link/free-ebook/978-1-83546-104-4

2. Submit your proof of purchase
3. That's it! We'll send your free PDF and other benefits to your email directly